Applied Nonlinear Optics

WILEY SERIES IN PURE AND APPLIED OPTICS

Advisory Editor

STANLEY S. BALLARD, University of Florida

Lasers, second edition, BELA A. LENGYEL

Introduction to Laser Physics, BELA A. LENGYEL

Laser Receivers, MONTE ROSS

The Middle Ultraviolet: its Science and Technology, A. E. S. GREEN, *Editor*

Optical Lasers in Electronics, EARL L. STEELE

Applied Optics, A Guide to Optical System Design/Volume 1, LEO LEVI

Laser Parameter Measurements Handbook, HARRY G. HEARD

Gas Lasers, ARNOLD L. BLOOM

Infrared System Engineering, R. D. HUDSON

Laser Communication Systems, WILLIAM K. PRATT

Optical Data Processing, A. R. SHULMAN

The Applications of Holography, H. J. CAULFIELD and S. LU

Far Infrared Spectroscopy, KARL D. MÖLLER and WALTER G. ROTHSCHILD

Optics, CHARLES S. WILLIAMS and ORVILLE A. BECKLUND

Applied Nonlinear Optics, FRITS ZERNIKE and JOHN E. MIDWINTER

Applied Nonlinear Optics

FRITS ZERNIKE   The Perkin–Elmer Corporation

AND

JOHN E. MIDWINTER   Post Office Research Department

A WILEY-INTERSCIENCE PUBLICATION

John Wiley & Sons New York/Sydney/Toronto/London

Library of Congress Cataloging in Publication Data:

Zernike, Frits, 1930–
  Applied nonlinear optics.

  (Wiley series in pure and applied optics)
  Bibliography: p.
  1. Nonlinear optics. I. Midwinter, John E.,
joint author. II. Title.

QC446.2.Z47   535   72-8369
ISBN 0-471-98212-1

Printed in the United States of America

10 9 8 7 6 5 4 3 2 1

Preface

This book is directed specifically to those physicists and engineers who are interested in the device applications made possible by the development of nonlinear optics over the past 10 years. It does not claim to cover the entire field; such subjects as higher order nonlinearities, stimulated Raman and Brillouin effect, and self-induced transparency are not included. These fields are a part of nonlinear optics, but so far their use in device applications seems to be limited and they are of interest mainly to the research physicist.

One subject does have definite device applications but is touched on here only superficially: the modulation of light by either the acoustooptic or the electrooptic effect. Enough work has been done in this field to fill a separate book, and rather than give an inadequate treatment, we prefer to leave the subject to someone more qualified.

The book is written entirely on a classical basis and no knowledge of quantum mechanics, other than the most elementary ideas, is required of the reader. However, a working knowledge of the elements of calculus and of electromagnetic theory is assumed.

The basic ideas of linear optics necessary for an understanding of the subject are covered in the first chapter. The theory of nonlinear optical interactions is introduced in Chapter 2; we provide sufficient detail to bring out the practical problems encountered but make no pretensions to absolute generality or rigor. The important subject of phase matching is treated in Chapter 3. Successful device applications of nonlinear optics are to a large extent dependent on the availability of the proper nonlinear materials. The subject of materials is treated in Chapter 4, and a comprehensive list of materials and their properties is given in an appendix. Since much materials research is still being conducted, it is virtually certain that this list will be incomplete at the time of publication. The remaining chapters are devoted to detailed treatments of second-harmonic generation, up-conversion, and parametric oscillation.

A few words should be said about the choice of dimensions. It appears that almost all the literature in the field uses the c.g.s. system. For this reason we have set aside our own preference for the MKS system and have used the c.g.s. system exclusively.

The authors wish to thank Professor Herbert F. Carleton and Cybex Incorporated for stimulating the original notes out of which this book has grown. One of us (JEM) is particularly grateful to Dr. T. P. McLean, whose Royal Radar Establishment Lecture Course affected his view of the subject and greatly influenced the presentation in this book. We also wish to thank Miss Andrea Miller, who patiently typed the manuscript through many revisions. Last but not least we wish to thank our wives and families, for their patience.

FRITS ZERNIKE
JOHN E. MIDWINTER

Norwalk, Connecticut
Ipswich, Suffolk, England
May 1972

Contents

Chapter 4 Nonlinear Materials   73

Applied Nonlinear Optics

1

Linear Optics: Wave Propagation in Anisotropic Materials

Throughout this book we work with nonlinearities of the refractive index of optical materials. We also make use of the anisotropy of the refractive index—either the natural anisotropy that occurs in certain materials or an anisotropy induced by outside influences. In this chapter we prepare the ground for these later developments by discussing the physical origin of the linear refractive index and the propagation of light in isotropic and anisotropic media.

1.1 LORENTZ MODEL

First we consider the classical model, due to Lorentz, of a single atom with one electron and a nucleus. If an electric field is applied, the distance between the electron and the nucleus changes: a polarization is induced. If the electric field is alternating, the polarization likewise alternates and, moreover, the frequency of the polarization is the same as that of the applied field. In other words, the electron now oscillates about its equilibrium position; it forms an oscillating dipole. This oscillating dipole radiates an electromagnetic wave. The frequency of that wave is the same as the frequency of oscillation of the dipole and therefore the same as the frequency of the applied field, but the phase of the wave is determined by the restoring force between the electron and the nucleus, as is the phase of the dipole.

Thus we have the following situation: part of the incident electromagnetic wave is used to drive the oscillating dipole, which in turn reradiates a wave with the same frequency, but with a different phase. The resultant wave, then, has a phase which is somewhere in between the phases of the incident wave and the reradiated wave.

We can do the same thing for a second atom with the resultant wave from the first one as the incident wave, and so on for N atoms in a row. Now take two parallel waves with the same initial phase and frequency and let one of these waves pass through the N atoms. At a point on the other side of the string of N atoms, the difference in phase between the two waves is of course proportional to N. A group of N atoms in a linear chain like this make up a material with thickness d, where d is proportional to N. Thus we arrive at the conclusion that the phase difference between the two waves is proportional to d. It appears as if the wave inside the material propagates more slowly than the one in a vacuum. The ratio between the two velocities is known as the absolute refractive index of the material:

$$n = \frac{c}{\text{velocity in the material}}$$

Often the refractive index given is the refractive index as measured with respect to air. Since the refractive index of air is approximately 1.00027, this number is slightly smaller than the absolute index.[126]

What we have tacitly assumed so far is that the light propagates only in the forward direction. However, the radiation pattern of an oscillating dipole has a $\sin \theta$ term* in it (it has a toroidal shape). Why then do the individual dipoles not radiate sideways, or, indeed, in the backward direction? The answer is that the phase of the radiation from every dipole is determined by the phase of the wave incident upon it. Therefore, even though each individual dipole does radiate in all directions, any radiation not in the forward direction is out of phase with the radiation in that same direction from other dipoles and does not build up. Only in the forward direction are all the dipoles phased correctly. They form a phased array of antennas. Thus the incident light is scattered coherently by the individual dipoles, but only in the forward direction is this scattering additive, giving rise to the refractive index. Later, we find instances where the phasing in the forward direction is not correct, and the result is no transmission.

In this simple model we have assumed that all the dipoles reradiate the same fraction of the incident wave. Only then does the radiation build up in the forward direction and interfere destructively in all other directions. If one dipole, or one group of dipoles, radiates a different fraction, some of this reradiated wave is visible in other directions. This phenomenon is normally called scattered radiation. Properly, it should be called incoherent scattering. The tendency to scatter is strong in materials with many impurities; in very pure materials, on the other hand, a beam of light is not visible in any but the forward direction.

Thus far we have assumed that all the energy used to drive the oscillating dipole is reradiated as an electromagnetic wave. Normally it is not just the

---

* θ is the angle between the axis of the dipole and the direction of observation.

electron which is made to oscillate, but the ensemble of electron and nucleus; moreover, there are interactions between the atoms. At certain frequencies part or all of the driving energy is not reradiated as an electromagnetic wave but causes vibrations of the atoms with respect to one another; it heats the material. In other words, part of the incident energy is absorbed.

To put all this on a more rigorous footing, we write the mathematical expression that describes the model mentioned previously. We consider the electron oscillating around its equilibrium position as a harmonic oscillator and write an equation of motion.[140]

$$\frac{d^2r}{dt^2} + 2\gamma \frac{dr}{dt} + \omega_0^2 r = -\frac{e}{m} E \tag{1.1}$$

Here r is the displacement of the electron from its equilibrium position; e is its charge, m its mass, and ω_0 its natural frequency; γ is a damping constant, and E is an applied electric field. We limit the analysis to motion in one dimension only.

We now consider an electric field

$$E = \mathscr{E} \cos (\omega t - \varphi)$$

However, writing the field this way produces, even in the linear case, a great deal of algebraic complication. To avoid this we use a complex number notation

$$E = E(\omega)e^{-i\omega t} + E^*(\omega)e^{+i\omega t} \tag{1.2}$$

where $E(\omega)$ is a complex amplitude that includes the phase

$$E(\omega) = \tfrac{1}{2}\mathscr{E}e^{i\varphi}$$

Here $\mathscr{E}$ is a real amplitude. Note that the convention of using script type to indicate specifically that a quantity is real is followed throughout the book. Note however, that Roman type does not always indicate that the quantity is complex.

We also have

$$E^*(\omega) = \tfrac{1}{2}\mathscr{E}e^{-i\varphi} = E(-\omega)$$

Substituting this in equation 1.1 gives a linear equation, which has as its solution:

$$r = -\frac{e}{m} E(\omega) \frac{e^{-i\omega t}}{\omega_0^2 - 2i\gamma\omega - \omega^2} + \text{complex conjugate} \tag{1.3}$$

A medium with an electron density N has a polarization density $P = -Ner$ or

$$P = \frac{Ne^2}{m} \frac{1}{\omega_0^2 - 2i\gamma\omega - \omega^2} E(\omega)e^{-i\omega t} + \text{complex conjugate} \tag{1.4}$$

and substituting

$$\chi(\omega) = \frac{Ne^2}{m} \frac{1}{\omega_0^2 - 2i\gamma\omega - \omega^2} \tag{1.5}$$

$$P = \chi(\omega)E(\omega)e^{-i\omega t} + \text{complex conjugate} \tag{1.6}$$

Equation 1.6 shows that the induced polarization is indeed proportional to the amplitude of the applied alternating field and has the same frequency. We now put this polarization as a source term in Maxwell's equations

$$\mathbf{\nabla} \times \mathbf{H} = \frac{1}{c} \frac{\partial \mathbf{D}}{\partial t} + \frac{4\pi}{c} \mathbf{j} \tag{1.7}$$

$$\mathbf{\nabla} \times \mathbf{E} = -\frac{1}{c} \frac{\partial}{\partial t} (\mu \mathbf{H}) \tag{1.8}$$

where

$$\mathbf{D} = \mathbf{E} + 4\pi \mathbf{P}$$

Equation 1.7 can be written

$$\mathbf{\nabla} \times \mathbf{H} = \frac{4\pi\sigma}{c} \mathbf{E} + \frac{1}{c} \frac{\partial \varepsilon \mathbf{E}}{\partial t}$$

where σ is the conductivity and $\varepsilon = (1 + 4\pi\chi)$.

We assume that the material is nonconducting and nonmagnetic; $\sigma = 0$ and $\mu = 1$, and we take the curl of both sides of equation 1.8. From the identity $\mathbf{A} \times \mathbf{B} \times \mathbf{C} = \mathbf{B}(\mathbf{A} \cdot \mathbf{C}) - \mathbf{C}(\mathbf{A} \cdot \mathbf{B})$ we know that $\mathbf{\nabla} \times \mathbf{\nabla} \times \mathbf{E} = \mathbf{\nabla}\mathbf{\nabla} \cdot \mathbf{E} - \mathbf{\nabla}^2\mathbf{E}$, and since $\mathbf{\nabla} \cdot \mathbf{E} = 0$, we find

$$\mathbf{\nabla}^2\mathbf{E} = \frac{\varepsilon}{c^2} \frac{\partial^2}{\partial t^2} \mathbf{E}$$

Restricting the problem to one dimension by setting $\partial/\partial x = \partial/\partial y = 0$, we have

$$\frac{d^2E}{dz^2} = \frac{\varepsilon}{c^2} \frac{\partial^2}{\partial t^2} E \tag{1.9}$$

As a solution of equation 1.9, we try a traveling wave

$$E(z, t) = \mathscr{E}e^{i(\omega t - kz)} + \text{complex conjugate} \tag{1.10}$$

which gives

$$k^2 = \frac{\varepsilon\omega^2}{c^2}$$

The propagation constant in the material, k, is equal to 2π times the number of waves per unit length. As such, k is a function of the frequency of the wave and of its velocity in the material, and since, as we have seen before, the velocity in the material is determined by the refractive index n, we have $k = n\omega/c$ and thus

$$n^2 = \varepsilon = 1 + 4\pi\chi \qquad (1.11)$$

Substituting equation 1.5 in equation 1.11, we find

$$n^2 = 1 + \frac{Ne^2}{m} \frac{4\pi}{\omega_0^2 - 2i\gamma\omega - \omega^2} \qquad (1.12)$$

For the case of $\gamma = 0$ (no damping), n is a real quantity, and its dependence on frequency is shown by the dotted curve in Figure 1.1. If $\gamma \neq 0$, n becomes complex. The imaginary part is a measure of the absorption and becomes large in the vicinity of ω_0. The real part follows the solid curve in Figure 1.1.

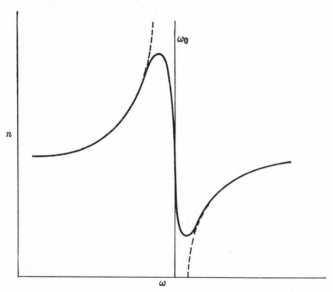

Figure 1.1 The behavior of the refractive index in the neighborhood of a natural frequency ω_0. The dotted curve represents no damping, the solid curve shows the influence of a damping factor.

In most materials there is more than one natural frequency ω_0 and therefore more than one absorption band. If the absorption is caused by the electron motion, the natural frequency is high, lying in the ultraviolet or in the visible region of the spectrum. If the frequency is the natural resonance of the vibrations of the atoms with respect to one another, then the frequency is lower and lies in the infrared. When several different groups of atoms are present in the crystal, each one has its own resonant frequency.* Thus the

* The infrared absorption bands are known as *reststrahlen* or residual rays.

infrared absorption spectrum gives direct information about the different atoms, or groups of atoms, in a material. Conversely, and of more importance to our subject, if the composition of a material is known, the spectral region in which it will be absorbing can be predicted. For example, many oxides have an absorption in a band centered at 9 μ. Thus oxides such as quartz and zinc oxide cannot be used as transparent materials with the 10.6-μ carbon dioxide laser.

The dispersion (the frequency dependence of the refractive index) always shows the same behavior. In between two absorption bands, the slope of the index versus frequency curve is always positive. At the high-frequency side, the index is high; it decreases relatively rapidly at first and then levels off until, in the vicinity of the next lower absorption band, it once again decreases more rapidly. Then, in the absorption band, the slope of the curve abruptly becomes negative: this is called anomalous dispersion, and the index increases sharply. On the low-frequency side, the slope once again becomes positive and the shape of the curve is analogous to the shape on the high-frequency side, but the entire curve is moved up and the refractive index is higher (Figure 1.2).

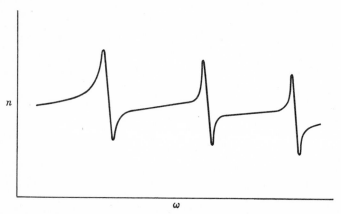

Figure 1.2   The general behavior of the refractive index as a function of frequency.

1.2 ANISOTROPY

It was mentioned earlier that the natural frequency ω_0, and therefore the refractive index, are influenced by the interaction between the atoms constituting the medium. It is quite conceivable, then, that in some materials this interaction between the atoms is not the same in all directions. Such media do indeed exist; they are called anisotropic. Most materials in which a natural anisotropy occurs are crystalline. We therefore limit the discussion to crystals.

In anisotropic crystals the dielectric constant is not a scalar, but a tensor. Since it relates the dielectric displacement in one direction to the fields in each of the three directions, this is a second-rank tensor. Thus we have*

$$D_i = \varepsilon_{ij}E_j \qquad (1.13)$$

We now assume that the expressions for the stored magnetic and electric energy density which are used in the isotropic case are still valid for anisotropic media. Furthermore, we assume that the energy flux is still given by the Poynting vector and that the electric field contains one half of the energy and the magnetic field the other half. The last two assumptions are, of course, also carried over from the isotropic case.

The stored electric energy density is

$$W_e = \frac{1}{8\pi}(\mathbf{E} \cdot \mathbf{D}) = \frac{1}{8\pi} E_j \varepsilon_{jk} E_k \qquad (1.14)$$

and so

$$\frac{\partial}{\partial t} W_e = \frac{1}{8\pi}\left(E_k \varepsilon_{kj} \frac{\partial E_j}{\partial t} + E_j \varepsilon_{jk} \frac{\partial E_k}{\partial t}\right) \qquad (1.15)$$

The Poynting vector is

$$\mathbf{S} = \frac{c}{4\pi}(\mathbf{E} \times \mathbf{H}) \qquad (1.16)$$

From equation 1.16 we find the net power flow into a unit volume

$$\mathbf{\nabla} \cdot \mathbf{S} = \frac{c}{4\pi}\mathbf{\nabla} \cdot (\mathbf{E} \times \mathbf{H})$$

Now by scalar multiplication of equation 1.7 by $\mathbf{E}$ and equation 1.8 by $\mathbf{H}$, and using the identity

$$\mathbf{\nabla} \cdot (\mathbf{A} \times \mathbf{B}) = \mathbf{A} \cdot (\mathbf{\nabla} \times \mathbf{B}) + \mathbf{B} \cdot (\mathbf{\nabla} \times \mathbf{A})$$

we find

$$\mathbf{\nabla} \cdot \mathbf{S} = \frac{\mathbf{E} \cdot (\partial \mathbf{D}/\partial t) + \mathbf{H} \cdot (\partial \mathbf{H}/\partial t)}{4\pi}$$

and using equation 1.13

$$\mathbf{\nabla} \cdot \mathbf{S} = \frac{1}{4\pi}\left(E_k \varepsilon_{k\ell} \frac{\partial E_\ell}{\partial t} + \mathbf{H} \frac{\partial \mathbf{H}}{\partial t}\right) \qquad (1.17)$$

* Here, as in the rest of the book, the Einstein convention of summation over repeated indices is followed (see Appendix 1).

The first term on the right is the electric energy density flux, and so from equation 1.15

$$\frac{1}{4\pi}\left(E_j\varepsilon_{jk}\frac{\partial E_k}{\partial t}\right) = \frac{1}{8\pi}\left(E_k\varepsilon_{kj}\frac{\partial E_j}{\partial t} + E_j\varepsilon_{jk}\frac{\partial E_k}{\partial t}\right)$$

it follows that $\varepsilon_{jk} = \varepsilon_{kj}$.

This means that the tensor is symmetrical; that is, it has only six independent components.

Writing out the expression 1.14 for the stored electric energy density, we have

$$W_e = \frac{1}{8\pi}\left(\varepsilon_{11}E_1{}^2 + \varepsilon_{22}E_2{}^2 + \varepsilon_{33}E_3{}^2 + 2\varepsilon_{23}E_2E_3 + 2\varepsilon_{13}E_1E_3 + 2\varepsilon_{12}E_1E_2\right)$$

$$(1.18)$$

By a suitable rotation of the coordinate axes, the last three terms can be eliminated, giving

$$W_e = \frac{1}{8\pi}\left(\varepsilon_x E_x{}^2 + \varepsilon_y E_y{}^2 + \varepsilon_z E_z{}^2\right) \tag{1.19}$$

where x, y, and z refer to the new axes. These new coordinate axes are called the principal dielectric axes. We now have

$$\begin{vmatrix} D_x \\ D_y \\ D_z \end{vmatrix} = \begin{vmatrix} \varepsilon_x & 0 & 0 \\ 0 & \varepsilon_y & 0 \\ 0 & 0 & \varepsilon_z \end{vmatrix} \begin{vmatrix} E_x \\ E_y \\ E_z \end{vmatrix} \tag{1.20}$$

1.3 WAVE PROPAGATION IN AN ANISOTROPIC CRYSTAL

Let us examine the transmission of a monochromatic plane wave through the crystal. We define the wave as in equation 1.10, except that the propagation constant is now a vector, often called the wave vector or the **k** vector.

$$\mathbf{k} = \frac{\omega n}{c}\mathbf{s}$$

where **s** is a unit vector normal to the wavefront.

For such a wave we can replace the operator ∇ by $(i\omega n/c)\mathbf{s}$ and $\partial/\partial t$ by $i\omega$. Then, assuming a nonconducting, nonmagnetic crystal, Maxwell's equations (1.7 and 1.8) become

$$\mathbf{H} \times \mathbf{s} = \frac{1}{n}\mathbf{D} \tag{1.21}$$

$$\mathbf{E} \times \mathbf{s} = -\left(\frac{1}{n}\right)\mathbf{H} \tag{1.22}$$

From equation 1.21 we see that **D** is perpendicular to **H** and **s**. From equation 1.22 it follows that **H** is perpendicular to **E** and **s**. Therefore, **D** and **H** constitute a proper transverse wave, and **D**, **s**, and **E** are all in a plane perpendicular to **H**. But (**E** × **H**), the Poynting vector, is not parallel to **s**; it lies in the same plane as **s**, **D**, and **E** but is perpendicular to **E**. This means that the direction of energy flow and the wave normal are not parallel (Figure 1.3).

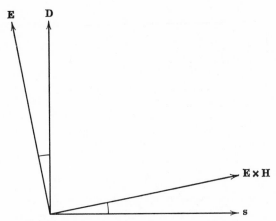

Figure 1.3   The angle between the wave normal and the ray direction. **H** is perpendicular to the plane of the figure, pointing upward.

Elimination of **H** from equations 1.21 and 1.22 gives

$$(\mathbf{E} \times \mathbf{s}) \times \mathbf{s} = -\left(\frac{1}{n^2}\right)\mathbf{D}$$

Again using the identity $\mathbf{A} \times \mathbf{B} \times \mathbf{C} = \mathbf{B}(\mathbf{A} \cdot \mathbf{C}) - \mathbf{C}(\mathbf{A} \cdot \mathbf{B})$

$$\mathbf{D} = (n^2)[\mathbf{E} - \mathbf{s}(\mathbf{s} \cdot \mathbf{E})]$$

By writing out the components of **D** and using equation 1.20 we obtain

$$D_x = n^2\left[\frac{D_x}{\varepsilon_x} - s_x(\mathbf{s} \cdot \mathbf{E})\right]$$

or

$$D_x = \frac{s_x[\mathbf{s} \cdot \mathbf{E}]}{1/\varepsilon_x - 1/n^2} \tag{1.23}$$

and forming the scalar product **D** · **s** we find, because of the orthogonality of **D** and **s**,

$$\frac{s_x^2}{1/n^2 - 1/\varepsilon_x} + \frac{s_y^2}{1/n^2 - 1/\varepsilon_y} + \frac{s_z^2}{1/n^2 - 1/\varepsilon_z} = 0 \tag{1.24}$$

Equation 1.24 is quadratic in n^2 and is known as Fresnel's equation. It has two independent solutions, $\pm n'$ and $\pm n''$. (The negative signs have no physical significance.) Accordingly there are two values, $\mathbf{D}'$ and $\mathbf{D}''$. To find their relative direction we use equation 1.23 to form the scalar product

$$\mathbf{D}' \cdot \mathbf{D}'' = (\mathbf{s} \cdot \mathbf{E})^2 \left\{ \frac{s_x{}^2}{[1/\varepsilon_x - 1/(n')^2][1/\varepsilon_y - 1/(n'')^2]} + \cdots + \text{etc.} \right\}$$

and this can be written

$$\mathbf{D}' \cdot \mathbf{D}'' = (\mathbf{s} \cdot \mathbf{E})^2 \frac{(n'n'')^2}{n'^2 - n''^2} \left[\frac{s_x{}^2}{1/\varepsilon_x - 1/(n')^2} \right.$$
$$\left. - \frac{s_x{}^2}{1/\varepsilon_x - 1/(n'')^2} + \frac{s_y{}^2}{1/\varepsilon_y - 1/(n')^2} - \cdots + \text{etc.} \right]$$

Since $(n')^2$ and $(n'')^2$ are solutions of Fresnel's equation 1.24, the expression in the brackets is zero and so

$$\mathbf{D}' \cdot \mathbf{D}'' = 0$$

That is to say, $\mathbf{D}'$ and $\mathbf{D}''$ are perpendicular to each other.

Thus we have found that an anisotropic crystal can transmit only waves that are plane polarized in one of two mutually orthogonal directions. We have learned that, in general, these polarizations "see" different refractive indices. As a rule, moreover, the direction of energy flow is not perpendicular to the wavefront.

One question that arises immediately is: what if the incident light is not polarized in either of the allowed directions? The obvious answer is that the incident light can always be decomposed into two beams linearly polarized in the allowed directions. In general, because of the different refractive indices, an incident beam that is linearly polarized in a direction other than either of the allowed directions will not be linearly polarized after transmission through the crystal.

1.4 THE INDEX ELLIPSOID

We also want to know how to find the allowed directions and the corresponding indices for an arbitrary direction of propagation. To obtain the answer, we go back to equation 1.19 and use equation 1.20. This gives

$$8\pi W_e = \frac{D_1{}^2}{\varepsilon_1} + \frac{D_2{}^2}{\varepsilon_2} + \frac{D_3{}^2}{\varepsilon_3}$$

If we substitute $D_1/\sqrt{8\pi W_e} = x$, etc., we find

$$\frac{x^2}{\varepsilon_1} + \frac{y^2}{\varepsilon_2} + \frac{z^2}{\varepsilon_3} = 1$$

and since $n = \sqrt{\varepsilon}$, we define the principal refractive indices by $n_i = \sqrt{\varepsilon}$ and write

$$\frac{x^2}{n_1^2} + \frac{y^2}{n_2^2} + \frac{z^2}{n_3^2} = 1 \qquad (1.25)$$

Equation 1.25 represents an ellipsoid with major axes in the x, y, and z directions. It is known as the index ellipsoid or the optical indicatrix, and it is used to find the two allowed directions of polarization and the indices for these directions. This is done as follows. Through the center of the ellipsoid we draw a plane perpendicular to the direction of propagation. The intersection of this plane and the ellipsoid is an ellipse. The two axes of this ellipse are parallel to the two directions of polarization, and the length of each axis is equal to twice the refractive index in that direction (Figure 1.4). Note that the direction of propagation which we have taken here is the *wave normal* or **k** vector, not necessarily the direction of energy flow.

We now examine how the indices of refraction vary when the direction of propagation is changed. Consider a plane through one of the polarization directions and the wave normal. The other polarization direction is of course perpendicular to this plane, and the intersection of this plane and the indicatrix is an ellipse. If we now rotate the wave normal in the plane through

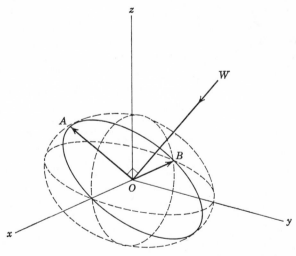

Figure 1.4   The indicatrix: OA and OB are the two allowed polarizations for the wave-normal W.

one of the polarization directions and the wave normal, the refractive index for one polarization varies elliptically. But the other refractive index is the axis of rotation, so to speak, and does not vary.

There are two kinds of optically anisotropic crystals. If all three axes of the indicatrix are unequal, the crystal is called biaxial, because in that case two optic axes can be defined (the definition of optic axis is given below). If the indicatrix is an ellipsoid of revolution (i.e., if two of the axes are equal), the crystal is uniaxial and the optic axis is perpendicular to the plane of the two equal axes. An optic axis is defined as that wave-normal direction in which the refractive index is independent of the direction of polarization, in other words, it is the direction perpendicular to a circular cross section of the indicatrix.

Now consider a biaxial crystal with $n_z > n_y > n_x$. In a plane parallel to x and y we have an ellipse with semiaxes n_x and n_y, but if we rotate the plane around the y axis, the minor axis of the ellipse becomes longer and eventually, after a 90° rotation, becomes equal to n_z. It is obvious that, at some point in between, there is an angle of rotation where the ellipse is a circle. And there are two such angles, one on each side of the x axis. There are thus two optical axes which are not parallel to the semiaxis of the ellipsoid (Figure 1.5). From elementary analytical geometry we can show that

$$\sin \theta = \frac{n_z}{n_y} \sqrt{(n_y^2 - n_x^2)/(n_z^2 - n_x^2)}$$

where θ is the angle between either optic axis and the z axis.

Whether a crystal is isotropic or anisotropic, and, in the latter case, whether it is uniaxial or biaxial, is determined by crystal symmetry. Thus crystals with a *cubic* symmetry are always *isotropic*. Crystals with the *trigonal*, *tetragonal*, and *hexagonal* symmetries are always *uniaxial;* and crystals with *orthorhombic*, *monoclinic*, and *triclinic* symmetries are always *biaxial.*

We shall first restrict our discussion to the most frequently encountered case of birefringence, that of uniaxial crystals. The indicatrix is then an ellipsoid of revolution. For the direction of polarization perpendicular to the optic axis, known as the *ordinary* direction, the index is independent of the direction of propagation. For the other direction of polarization, known as the *extraordinary* direction, the index changes elliptically between the value of the ordinary index n_o, when the wave normal is parallel to the optic axis and the extraordinary index n_e, when the wave normal is perpendicular to the optic axis. The two beams of light so produced are often referred to as *o*-rays and *e*-rays, respectively. When the wave normal is in a direction θ to the optic axis, the extraordinary index is given by

$$n(\theta) = \frac{n_e n_o}{(n_o^2 \sin^2 \theta + n_e^2 \cos^2 \theta)^{1/2}} \tag{1.26}$$

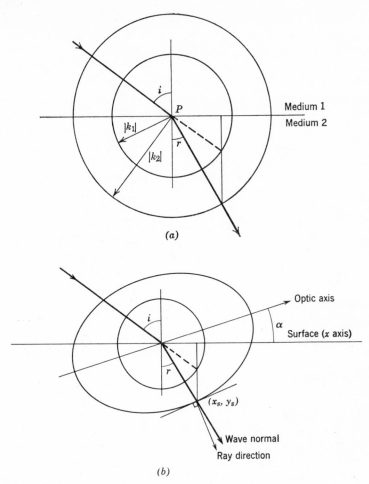

Figure 1.5 The wave vector constructions for refraction: (*a*) at the surface of an isotropic material, (*b*) at the surface of an anisotropic material.

If the value of $n_e - n_o$ is larger than zero, the birefringence is said to be positive, and for $n_e - n_o$ smaller than zero the birefringence is negative; the corresponding crystals are called positive or negative uniaxial.

1.5 REFRACTION AT THE SURFACE OF AN ANISOTROPIC CRYSTAL

For isotropic materials the refraction of a ray incident on the surface at an angle i to the normal is given by Snell's law: $\sin i = n \sin r$, where r is the angle between the normal and the refracted ray. This law remains valid for anisotropic materials, but for an e-ray it should properly be written as $\sin i = n(r) \sin r$, indicating that the index is dependent on the value of r.

In most textbooks the refraction of an e-ray at the surface is treated using Huygens's construction.[82] We employ a different method—the $\mathbf{k}$ vector construction—however, because of its usefulness, especially in later parts of this book. The vector $\mathbf{k} = k\mathbf{u}$ where $\mathbf{u}$ is a unit vector in the direction of propagation represents the momentum of the photon. This momentum has to be conserved, and so we find that the component of $\mathbf{k}$ parallel to the boundary has to be continuous across that boundary.

First let us look at the isotropic case (Figure 1.5a). To find the angle r for a ray incident at an angle i in a point P on the boundary between two isotropic media with refractive indices n_1 and n_2, we draw two concentric circles with radii $|\mathbf{k}_1|$ and $|\mathbf{k}_2|$ and P as the center. Here $\mathbf{k}_1$ is the wave vector in medium 1 and $\mathbf{k}_2$ is the wave vector in medium 2. Note that $|\mathbf{k}_1|/|\mathbf{k}_2| = n_1/n_2$. The direction of the refracted ray is found by continuing the incident ray until it intersects the circle with radius $|\mathbf{k}_1|$ in medium 2, and then drawing a line normal to the boundary through this intersection. The intersection of this normal with the circle with radius $|\mathbf{k}_2|$ gives the point at which the refracted ray intersects this circle.

It is easily seen that the angle r thus obtained satisfies Snell's law. It is also clear that we have indeed used the condition that the $\mathbf{k}$ vector component parallel to the boundary be continuous across the boundary.

Applying the same condition when medium 2 is anisotropic, we draw the ellipse which is the locus of the ends of the wave vectors in the different directions in medium 2 and repeat the operation performed previously (Figure 1.5b). The direction we find in this way is the wave normal (i.e., the direction normal to $\mathbf{D}$). That this is so can be understood by realizing that the $\mathbf{k}$ vector gives the phase propagation and thus has to be normal to the wavefront. It can be shown that the ray direction (normal to $\mathbf{E}$) is parallel to the normal to the ellipse at the point of intersection.

Note that the ellipse drawn in Figure 1.5b is the locus of the wave vector; it is not a section of the indicatrix, nor is it a section of the surface at which light emanating from P has arrived after a given time interval. (This surface is variously known as the wave surface or the ray surface.) The latter two sections have the same shape, but they form ellipses at right angles to the one used in Figure 1.5b.

We now find an expression for the angle r as a function of i and α (Figure 1.5b). Setting $1/k_e = e$ and $1/k_o = o$, the equation for the ellipse is

$$e^2 x^2 + o^2 y^2 = 1 \qquad (1.27)$$

Rotating the coordinates over an angle $-\alpha$ so that the x axis becomes parallel to the surface, we obtain from equation 1.27

$$(o^2 \cos^2 \alpha + e^2 \sin^2 \alpha)x^2 + (o^2 \sin^2 \alpha + e^2 \cos^2 \alpha)y^2$$
$$+ (e^2 - o^2)(\sin 2\alpha)xy = 1 \quad (1.28)$$

and cotan $r = -y_s/x_s$, where (x_s, y_s) are the coordinates of the point of intersection with respect to the new axes. Now $x_s = \sin i$. Substituting this in equation 1.28 we find for a positive crystal

$$\text{cotan } r = \frac{(e^2 - o^2) \sin 2\alpha + 2\sqrt{o^2 \sin^2 \alpha/\sin^2 i + e^2 \cos^2 \alpha/\sin^2 i - o^2 e^2}}{2(o^2 \sin^2 \alpha + e^2 \cos^2 \alpha)}$$

Here the plus sign is used when the incident ray and the optic axis are on opposite sides of the normal to the surface; the minus sign is used when both are on the same side. For a negative crystal, o and e and the plus and minus signs should be interchanged.

If θ is the angle between the wave normal and the optic axis and ρ is the angle between the ray direction and the optic axis, we have

$$\tan \rho = \left(\frac{n_o}{n_e}\right)^2 \tan \theta \quad (1.29)$$

1.6 APPLICATIONS OF BIREFRINGENCE

An important application of birefringence in nonlinear optics makes use of the fact that the normal color dispersion of the material may be compensated with the birefringence. In other words, two waves with different frequencies ω_1 and ω_2 may be made to propagate with identical velocities inside the crystal by making one an o-wave and the other an e-wave, and by choosing the angle of propagation θ so as to obtain $n_{\omega_1}{}^o = n_{\omega_2}{}^e(\theta)$. This does *not* mean that the ray directions of the two waves will be parallel (see also Chapter 3).

Apart from this, birefringent crystals serve in a number of ways in optical technology. Their major applications are in components that change the state of polarization of a light beam. Consider a plane-parallel slab of a uniaxial birefringent crystal (e.g., ADP, ammonium dihydrogen phosphate) cut with the optic axis parallel to its face. A beam of light linearly polarized in a direction 45° to the optic axis of the crystal is incident normal to the face. Inside the crystal there are now two linearly polarized waves, one polarized parallel to the optic axis and the other perpendicular to it. These two waves have equal amplitudes, $1/\sqrt{2}$ as large as the amplitude of the incoming wave (neglecting surface reflections). The extraordinary wave has a wavelength

λ/n_e inside the crystal; the ordinary wave has a wavelength λ/n_o. If the thickness of the crystal is d, there are $n_e d/\lambda$ extraordinary waves and $n_o d/\lambda$ ordinary waves in the crystal. If $(n_e - n_o)d = \lambda/2$, there is a 180° phase difference between the two polarization components at the exit face of the crystal, and so the resultant of these two components is polarized at right angles to the polarization of the incident beam. It is easy to see that by choosing angles other than 45°, such a *half-wave plate* may be used to rotate the polarization direction of a plane-polarized beam over any desired angle. If $(n_e - n_o)d = \lambda/4$, the plate is known as a *quarter-wave plate* and may be used to produce circularly or elliptically polarized light.

An incident beam linearly polarized at 45° to the axis of a quarter-wave plate comes out circularly polarized and, after being reflected back through the plate, it comes out linearly polarized but with the perpendicular direction of polarization. Thus a quarter-wave plate can be used as an isolator. However, this arrangement works only if the state of polarization of the circularly polarized light is not changed by the components to be isolated.

1.7 ORIENTATION OF THE CRYSTAL

The possibility of a change in the state of polarization of a light beam transmitted through an anisotropic crystal, discussed in the preceding section, leads to a method of finding the direction of the optic axis in a uniaxial crystal. If a relatively thick slab of the crystal is put in parallel light between crossed polarizers, the state of polarization of the light transmitted by the polarizers remains unchanged only if the optic axis of the crystal is either parallel or perpendicular to the direction of polarization of the polarizer. At any point in between, the crystal produces either circularly or elliptically polarized light, and this is transmitted by the analyzer. In practical terms, if the optic axis is in or near a plane parallel to the plane of the polarizer and the analyzer, rotation of the crystal produces two mutually perpendicular positions where the crystal appears black. In one of these positions, the direction of the optic axis is parallel to the direction of polarization of the light transmitted by the polarizer.

The degree of parallelism can be determined very accurately, but this information, of course, is of little value unless it is possible to ascertain which one of the two directions is indeed the optic axis. To do that, the crystal is again placed between crossed polarizers, but this time in strongly convergent (or divergent) light. If the optic axis is parallel to the direction of observation, a set of concentric colored rings is seen, with a black cross through the center. Thus it can be determined which of the two previous directions is the axis.

The explanation of the colored rings is not given here, since it can be found in standard textbooks on optics.[82] Using the method described, the axis can be determined to within 1°. Similar methods may be used for biaxial crystals. Once the axis has been found to within 1°, flat and parallel faces can be put on which are perpendicular to the axis to within the measured accuracy. The exact direction can then be determined to any accuracy, limited only by the diffraction of the size of the beam used for the measurement. To do this the crystal is placed on a rotary table between crossed polarizers in collimated light. The axis of rotation of the table should be at approximately 45° to the direction of polarization of the polarizer.

When the optic axis is exactly perpendicular to the entrance face, the field is dark and remains dark when polarizer and analyzer are rotated together in the crossed position. In general, however, the field ceases to be dark when the crystal is inserted. A small rotation of the table in either direction serves to restore darkness, but in only one of these positions is the light transmitted parallel to the optic axis. To determine which position is the optic axis, it is necessary to again rotate the polarizer and the analyzer together in the crossed position. If the field does not remain dark, the light is transmitted parallel to the direction which gives a circular fringe and not parallel to the optic axis. Once the direction of the axis is found the crystal should be turned 90° in the plane perpendicular to the light and the procedure repeated. It should be remembered that, to determine the direction of the optic axis, the angle between the normal to the entrance face and the position in which the light is transmitted parallel to the optic axis should be reduced to the angle inside the crystal by using Snell's law and the ordinary index.

In this way the direction of the optic axis can be found by purely optical methods. In nonlinear optics, however, we also need to know the direction of the other axes in the crystal: the x axis and the y axis. These cannot be found by a simple optical method.

An often-used method is x-ray orientation,[167] but such other methods as determination of the shape of etch-pits on different faces,[163] and the piezo-electric effect[130] are also appropriate. A crude determination of the optical axis direction is always useful, however, since any one of these methods becomes much easier to use once a face normal to the axis has been put on the crystal. Of course, if the crystal is available in its entire form, with natural crystal faces, the axes can be determined from these. This is often the case with solution-grown crystals, such as ADP and KDP (potassium dihydrogen phosphate).

1.8 BIAXIAL CRYSTALS

Thus far we have restricted ourselves to uniaxial crystals, where the index ellipsoid is an ellipsoid of revolution. We now consider the case of biaxial crystals, where the index ellipsoid has three unequal axes. There are three principal refractive indices in the crystal n_x, n_y, and n_z. We use the normal convention that the principal refractive indices are in the order $n_z > n_y > n_x$.

The difference between uniaxial and biaxial crystals becomes clearly apparent when we consider the **k** vector surfaces (i.e., the locus of the end of the **k** vector as a function of direction). Since for any anisotropic crystal there are two indices for the two orthogonal polarizations, the wave vectors always define two surfaces. In the case of a uniaxial crystal, one of the surfaces, that for the ordinary direction of polarization, is of course a sphere. The other surface is an ellipsoid of revolution. The intersection of this surface with a plane was used in Section 1.5. Note again that this surface is not the same ellipsoid of revolution as the indicatrix. For example, for a positive uniaxial crystal, the z axis of the indicatrix is the longer axis, whereas for the wave vector surface the z axis for the same case is the shorter axis.

For biaxial crystals neither surface is spherical. Let us first examine the intersections of the two surfaces with the principal planes (i.e., the x-y, the y-z, and the x-z planes) in Figure 1.6. In each plot the cross sections form an

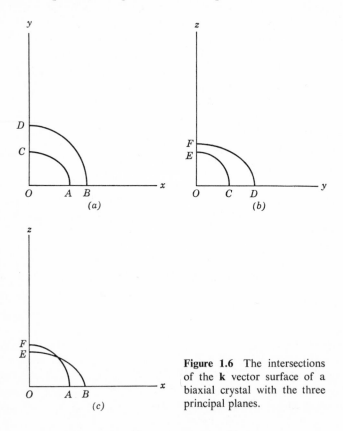

Figure 1.6 The intersections of the **k** vector surface of a biaxial crystal with the three principal planes.

ellipse and a circle. The polarization is always perpendicular to the plane of the figure for the circle; for the ellipse it is in the plane of the figure and is always tangential to the ellipse at every point.

For example, in Figure 1.6a the polarization at A is parallel to the y axis, but at C it is parallel to the x axis. In Figure 1.6c the ellipse and the circle cross. The complete surface, which appears in Figure 1.7, consists of two

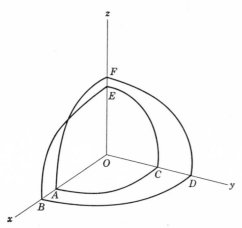

Figure 1.7 The two-sheeted **k** vector surface for a biaxial crystal.

sheets that intersect in the x-y plane. Obviously the point of intersection is the point where the optic axis intersects the surface.

In Figure 1.8a and b we present the two surfaces separately and we show the allowed polarization as a set of contour lines on each one of these surfaces. The allowed polarization at a point on one of these contour lines is tangent to the line in that point. Clearly at any arbitrary point on either surface, the allowed polarization will not be parallel to any of the principal planes unless the point itself lies in one of the principal planes. If we take two points, one on each of the surfaces, that lie on a line through the origin, the polarization in one of the points will be at right angles to the polarization in the other point.

For completeness we add here a note about the direction of the coordinate axes. To obtain equation 1.19 from equation 1.18, we performed a rotation of the coordinate axes. The exact angles through which we rotated were determined by the values of the components ε_{ij} of the dielectric tensor. However, just as in the isotropic case, there will be a dispersion with frequency in these components, and so the rotation could be frequency dependent. Indeed, when the direction of the principal axes is not given by crystal symmetry, there is a dispersion in the axis directions. In monoclinic crystals this

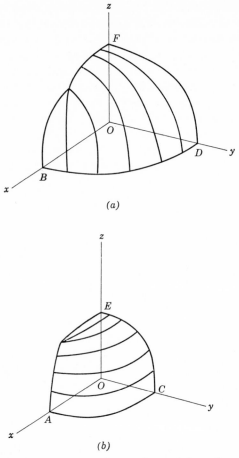

Figure 1.8 Polarization contours on the two sheets of the surface of Figure 1.7.

dispersion exists for only two of the axes; in triclinic crystals all three axes have a frequency dispersion. In orthorhombic crystals the three axes are given by the crystal symmetry and are therefore independent of frequency.

1.9   OPTICAL ACTIVITY

From the foregoing discussion, we would expect that light traveling parallel to the optic axis would be propagated without change in polarization. In some substances, however, this is not true. If a slab of quartz cut with the optic axis normal to the surface is placed between crossed polarizers in monochromatic light, some light is almost always transmitted. Only rotation of the analyzer will restore full extinction. This phenomenon is known as optical activity: the plane of polarization of the light is rotated inside the crystal, the exact amount of rotation necessary for extinction depends on the wavelength.

In a crystal that shows natural optical activity, the direction in which the polarization is rotated is independent of the direction in which the light is transmitted through the crystal. It follows that if a beam going from left to right is reflected and made to go back along the same path through the crystal, its polarization will again be parallel to the original direction of polarization. Thus this effect cannot be used to make optical isolators. A thorough description of optical activity is given in the book by Nye, recommended at the end of the chapter.

1.10 INDUCED ANISOTROPY

In our discussion of the natural optical anisotropy in crystals, it was mentioned that the anisotropy of the dielectric constant occurs because the different atoms in the medium do not "see" the same electric field in all directions. It is not surprising then, that an anisotropy can be induced in an isotropic material by an outside influence, such as a strain or an electric field. To be observable however, the outside influence must be large enough to perturb the interatomic fields to a significant extent. In other words, the effects are small—they can be observed readily only because small index variations are detectable by interferometric methods. The induced anisotropy produces a differential change of the refractive indices for two orthogonally polarized beams, and the resulting change in the interference of these beams is easily observed (e.g., between crossed polarizers).

Obviously these effects can be used to modulate a beam of light. A sound wave can interact with a light wave, an electric field can be used to switch the polarization of a light beam, or even light beams can modulate each other.

1.11 ELECTROOPTIC EFFECT

Before going deeper into the nonlinearities of the polarizability we discuss the electrooptic effect, which illustrates many points to be used again later. We treat the most general case, that is, the first-order electrooptic effect in anisotropic crystals.[16,130]

As we have seen, the dielectric constant is a second-rank tensor, and the indicatrix, when not referred to its principal axes, is given by

$$b_{11}x^2 + b_{22}y^2 + b_{33}z^2 + 2b_{23}yz + 2b_{13}xz + 2b_{12}xy = 1 \qquad (1.30)$$

where $b_{11} = 1/n_{11}^2$, etc.

The electrooptic coefficient gives the change in the dielectric constant as a result of an applied field; this coefficient is again a tensor, but since it operates on a second rank tensor, it has to be a tensor of the third rank. In general, a third-rank tensor has 27 independent components, but because the second-rank tensor on which we operate is symmetrical (only six independent components), the electrooptic tensor has at most only 18 independent elements. Moreover, the tensor must be invariant to those symmetry operations which transform the crystal into itself. In most cases this condition means that many of the 18 components are zero. In particular, if the crystal has a center of symmetry, the effect has to be the same for E as for $-E$, and, obviously, this can be so only if all components are zero. In other words, *centrosymmetric crystals do not show a first-order electrooptic effect.*

The existence of only 18 independent elements permits a simplification in notation. We contract the indices in equation 1.30 according to the following scheme:

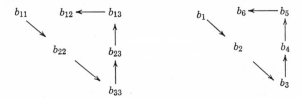

The arrows in the figures indicate the sequence. Using this contraction, the indicatrix becomes

$$b_1 x^2 + b_2 y^2 + b_3 z^2 + 2b_4 yz + 2b_5 xz + 2b_6 xy = 1$$

The electrooptic tensor gives the change of b_i as a result of an applied field $\mathscr{E}_j$. This change is

$$\Delta b_i = r_{ij} \mathscr{E}_j$$

where i runs from 1 through 6 and j runs from 1 through 3. The electrooptic tensor, then, can be represented by a matrix with 6 rows and 3 columns. Remember, however, that the tensor is really a third-rank tensor and transforms as such!

Since the electrooptic effect produces a change in index, the changed indices can still be represented by an indicatrix. If we start with the indicatrix as referred to its principal axes, the application of an electric field will, in general, produce off-diagonal terms. The new indicatrix then will not be referred to its principal axes, but a new set of principal axes can be obtained by an appropriate rotation of the coordinate system.

To illustrate this we consider the case of ammonium dihydrogen phosphate (ADP). The matrix has only two independent components, $r_{41} = r_{25}$

and r_{63}. We write the indicatrix of the crystal without applied field as follows:

$$o^2 x^2 + o^2 y^2 + e^2 z^2 = 1 \tag{1.31}$$

where $o = 1/n_o$ and $e = 1/n_e$. A field $\mathscr{E}_x$ applied in the x direction will add a term $2r_{41}\mathscr{E}_x yz$; to equation 1.31; similarly, $\mathscr{E}_y$ gives $2r_{41}\mathscr{E}_y xz$ and $\mathscr{E}_z$ gives $2r_{63}\mathscr{E}_z xy$.

We take the case of a field in the z direction. Equation 1.31 then becomes

$$o^2 x^2 + o^2 y^2 + e^2 x^2 + 2r_{63}\mathscr{E}_z xy = 1 \tag{1.32}$$

Rotation of the coordinate system around an angle α gives

$$xy = \frac{(y'^2 - x'^2)\sin 2\alpha}{2} + x'y'\cos 2\alpha \tag{1.33}$$

where the term on the left refers to the old system and the right-hand side refers to the coordinate system after rotation. Substituting equation 1.33 in equation 1.32 and setting $\alpha = 45°$, we find

$$(o^2 - r_{63}\mathscr{E}_z)x'^2 + (o^2 + r_{63}\mathscr{E}_z)y'^2 + e^2 z'^2 = 1 \tag{1.34}$$

Equation 1.34 shows that the crystal has become biaxial. The principal axes of equation 1.34 are rotated 45° around the z axis by comparison with those of equation 1.31. Note that the rotation is independent of voltage. This might seem inconsistent: a sudden rotation over 45° when the field is switched from zero to a very small value. We quickly realize, however, that the rotation is only a way of finding the principal axes of the indicatrix, and at zero field the x and y axes of the indicatrix could equally well be picked at 45° to the crystallographic x and y axes.

From equation 1.34 we see that the biaxial crystal has the following principal refractive indices:

$$n_x = n_o(1 + \tfrac{1}{2}n_o^2 r_{63}\mathscr{E}_z)$$
$$n_y = n_o(1 - \tfrac{1}{2}n_o^2 r_{63}\mathscr{E}_z)$$
$$n_z = n_e$$

where x, y, z refer to the coordinate system after rotation.

An application of this interaction is as a shutter or a modulator. Suppose that a plane-polarized beam is transmitted through the ADP crystal in a direction parallel to the z axis. Without the applied field, the transmission would be independent of the direction of polarization, but when the field is turned on, different directions of polarization "see" a different refractive index. Specifically, if the incident beam is polarized parallel to either the crystallographic x axis or the crystallographic y axis, and if the applied voltage is such that

$$n_o{}^3 r_{63} \mathscr{E}_z d = \frac{\lambda}{2} \qquad (1.35)$$

the crystal acts as a half-wave plate with its axes at 45° to the direction of polarization, rotating the plane of polarization over 90°. A subsequent polarizer then blocks this direction of polarization.

In the configuration just described the electric field is applied parallel to the light beam; this means that transparent electrodes are needed. It also means that the induced retardation is independent of the thickness of the plate because, since $\mathscr{E}_z = V_z/d$, where V is the applied voltage, equation 1.35 becomes

$$n_o{}^3 r_{63} V_z = \frac{\lambda}{2}$$

The voltage necessary to obtain a half-wave retardation is known as the half-wave voltage.

The following books are recommended for further reading on the subjects covered in this chapter:

J. F. Nye, *Physical Properties of Crystals*, Oxford, Clarendon Press, 1960.
M. Born and E. Wolf, *Principles of Optics*, London, Pergamon Press, 1959.

2

Nonlinear Optics

2.1 INTRODUCTION

In the preceding chapter it was shown that the refractive index of a material results from the polarization of that material by the electric field of the transmitted radiation. This polarization was completely linear: for a field increase of a factor of 2, the polarization increased by the same factor. But we know from other areas of physics that the linear dependence of one physical quantity on another is almost always an approximation, having validity in a certain limited range only. Perhaps the most familiar example of this is Hooke's law and its breakdown for large enough stresses. So we expect that the polarization, also, will be linear only for limited values of the field strength. For example, the electrooptic effect, which we introduced as the change in refractive index due to the application of an electric field, is really observable only because the electric field is large enough for the nonlinearity of the polarization to show up.

We chose to use a temporally uniform field before, but clearly an alternating field would have similar consequences. However, if the refractive index of a crystal is modulated by such a field, of frequency ω_2, then a beam with frequency ω_1 passing through the crystal will be phase modulated, and this phase modulation will give rise to side bands at combination frequencies, such as the sum frequency and the difference frequency. Thus we can create different frequencies by modulating the parameters of the crystal. This is why these interactions are often called parametric frequency conversion.

The alternating field at ω_2 also modulates the refractive index which it itself "sees," giving rise to a side band, a harmonic overtone, at $2\omega_2$. The same thing happens for the beam at ω_1, giving a side band at $2\omega_1$. However, as mentioned, the nonlinearities of the refractive index are so small that generally they can be detected only when the amplitude of the electric field is of the same order of magnitude as the interatomic fields. Thus the side band at $2\omega_1$ is observable only if the beam at ω_1 has a high intensity. But, clearly, even if the beam at ω_1 is of low intensity, a sum frequency, at $\omega_1 + \omega_2$, or a difference frequency, at $\omega_1 - \omega_2$, may still be created, as long as the intensity of the beam at ω_2 is large.

Thus the nonlinearity of the polarization can be used to "detect" a weak signal, at a wavelength for which sensitive detectors do not exist, by transforming its frequency to the visible range, where it can be detected sensitively with a photomultiplier. This is called parametric up-conversion (see Chapter 6).

The energy balance in some of the possible interactions is such that the energy from one frequency (the "pump") can be fed into two lower frequencies. If the pump is intense enough to overcome losses, neither of the lower frequency signals has to be provided from the outside, but they can be made to build up out of the noise. In this way a tunable source can be obtained. This is the parametric oscillator (see Chapter 7).

Although these interactions are reminiscent of the parametric interactions that have been used for years in the microwave field, there is one significant difference: local nonlinearities, such as exist in junctions, are used in microwaves; in nonlinear optics the interaction takes place in the bulk.

In this chapter we introduce the nonlinearity of the polarization which is responsible for these interactions. To keep the discussion simple, we are at first concerned with the generation process only; what happens to the incident radiation is deferred until later. This is not a physically unrealistic situation, in many cases the amount of radiation generated is very small, and the influence of the interaction on the incident radiation is completely negligible.

The electrooptic effect, for example, can be seen as a frequency-mixing interaction between the incident radiation and a field caused by an externally applied voltage. This interaction produces a very weak beam, 90° out of phase with the incident radiation, which shows up as a phase shift in the transmitted radiation only because it has the same frequency as the incident beam. However, under ordinary circumstances, the influence of this interaction on the applied voltage is negligible.

2.2 NONLINEARITIES OF THE POLARIZATION

Rather than treating the interaction as a modulation of the refractive index, as we did previously, it is more convenient to consider it as a result of the nonlinearity of the polarization. Thus we include nonlinear terms in the polarization and write

$$P = \alpha E(1 + a_1 E + a_2 E^2 + a_3 E^3 + \cdots)$$

where α is the linear polarizability and a_1, a_2, etc., are nonlinearities of increasingly high order. Again, as in the electrooptic effect, we see immediately that for crystals with a center of symmetry, a_1, a_3, etc., have to be identically equal to zero, since the symmetry requires that $a_1 E = -a_1 E$, etc. Concentrating only on the first nonlinearity (i.e., $a_1 E$), we can write

$$\mathscr{P} = 2d\mathscr{E}^2 \tag{2.1}$$

where $\mathscr{P}$ is the nonlinear polarization due to this first nonlinearity only.*

Now let us suppose that we want to examine the interaction in the crystal of two traveling waves†

$$\mathscr{E}_1(z, t) = \mathscr{E}_1 \cos (\omega_1 t + k_1 z)$$

and

$$\mathscr{E}_2(z, t) = \mathscr{E}_2 \cos (\omega_2 t + k_2 z)$$

To do this we substitute the superposition of these two waves for $\mathscr{E}$ in equation 2.1. This gives

$$\mathscr{P} = 2d[\mathscr{E}_1{}^2 \cos^2 (\omega_1 t + k_1 z) + \mathscr{E}_2{}^2 \cos^2 (\omega_2 t + k_2 z) \\ + 2\mathscr{E}_1\mathscr{E}_2 \cos (\omega_1 t + k_1 z) \cos (\omega_2 t + k_2 z)]$$

By using the elementary trigonometric relations $\cos^2 \alpha = (1 - \cos 2\alpha)/2$ and $\cos \alpha \cos \beta = [\cos (\alpha + \beta) + \cos (\alpha - \beta)]/2$, we next find that the polarization consists of a number of components with different frequencies; that is,

$$\mathscr{P}_{2\omega_1} = d\mathscr{E}_1{}^2 \cos [2(\omega_1 t + k_1 z)]$$

$$\mathscr{P}_{2\omega_2} = d\mathscr{E}_2{}^2 \cos [2(\omega_2 t + k_2 z)]$$

$$\mathscr{P}_{\omega_1+\omega_2} = 2d\mathscr{E}_1\mathscr{E}_2 \cos [(\omega_1 + \omega_2)t + (k_1 + k_2)z] \tag{2.2}$$

$$\mathscr{P}_{\omega_1-\omega_2} = 2d\mathscr{E}_1\mathscr{E}_2 \cos [(\omega_1 - \omega_2)t + (k_1 - k_2)z]$$

and a steady term

$$\mathscr{P}_{\text{direct}} = d(\mathscr{E}_1{}^2 + \mathscr{E}_2{}^2) \tag{2.3}$$

* We use $2d\mathscr{E}^2$ rather than $d\mathscr{E}^2$ in order to make d conform to accepted practice.

† In general, these two waves do have different phases and, strictly taken, they should be written as follows:

$$\mathscr{E}_1(z, t, \varphi_1) = \mathscr{E}_1(\cos \omega_1 t + k_1 z + \varphi_1)$$

and

$$\mathscr{E}_2(z, t, \varphi_2) = \mathscr{E}_2(\cos \omega_2 t + k_2 z + \varphi_2)$$

For simplicity we have omitted the phases φ_1 and φ_2 here.

Thus we see that the nonlinear polarization contains a steady term, a sum- and a difference frequency, and the first overtone of both input frequencies. These overtones are normally called the second harmonics.

Note the factors of 2 in the expressions for the sum and difference frequencies.

In the linear case we found that the polarization wave generated an electromagnetic wave of the same frequency as the incident wave, and that, providing there was no absorption, the fraction of the incident energy that caused the polarization wave was reradiated without any loss. Now we find that the different components of the nonlinear polarization similarly generate electromagnetic waves, but these electromagnetic waves have frequencies different from those of the incident waves. As a result, the fraction of the incident energy used to create the nonlinear polarization can be reradiated at one or more of a number of different frequencies.

To find out which one of these frequencies will be radiated we have to examine the phases of the different waves involved. Again, in the linear case we found that the different radiating dipoles form a phased array of antennas, and that, because of this, they radiate a wave in the forward direction. However, the phase of the radiating dipole at any point in the nonlinear medium is not governed by the phase of a single wave at the same frequency, but by the relative phases at that point of two waves of differing frequencies. Moreover, the electromagnetic wave that is radiated by this dipole has a propagation velocity determined by its own frequency and by the refractive index of the material for that frequency.

In general, because of normal dispersion, this propagation velocity is different from the velocity with which the polarization wave is propagated; the electromagnetic wave radiated by a dipole at one point no longer couples to the energy radiated by a similar dipole at some other point, and destructive interference will occur. In other words, the entire system of radiating dipoles does not necessarily form a correctly phased array of antennas any more, as it did in the linear case. For example, equation 2.2 shows that the sum frequency component of the nonlinear polarization has a propagation constant $(k_1 + k_2)$. Generally this propagation constant is not equal to $k_3 = n_3\omega_3/c$, the propagation constant at $\omega_3 = \omega_1 + \omega_2$.

A technique, known as phase matching, exists for restoring the proper phasing of the dipole. It involves precise control of the indices at the three frequencies involved in the mixing process, to match the velocities of propagation of the polarization wave and the electromagnetic wave which it radiates. As a rule, it is possible to do this with only one of the frequency components of the polarization wave, and this is the frequency that will be generated efficiently (see Chapter 3).

Before beginning our discussion of the electromagnetic radiation generated by the nonlinear polarization, we first take a closer look at the origin of the nonlinear terms.

2.3 THE ANHARMONIC OSCILLATOR

To obtain the nonlinearity of the polarization, we introduce an anharmonic term into the Lorentz model of Chapter 1. Instead of equation 1.1, we write[23]

$$\frac{d^2r}{dt^2} + 2\gamma \frac{dr}{dt} + \omega_0^2 r - \xi r^2 = -\frac{e}{m} E \qquad (2.4)$$

Because of this anharmonic term, equation 2.4 cannot be solved in the same simple way that served for equation 1.1, and in fact an exact solution would be quite complicated. However, since the anharmonic term is very small compared with the harmonic one, its effects are also likely to be small compared with the linear field dependence of r. We therefore try a solution in the form of a power series

$$r = r_1 + r_2 + r_3 + \cdots + \text{etc.} \qquad (2.5)$$

where

$$r_\ell = a_\ell E^\ell \qquad (2.6)$$

Putting equation 2.5 in equation 2.4 and collecting terms of the same order in E, we find

$$\frac{d^2r_1}{dt^2} + 2\gamma \frac{dr_1}{dt} + \omega_0^2 r_1 = -\frac{e}{m} E \qquad (2.7)$$

$$\frac{d^2r_2}{dt^2} + 2\gamma \frac{dr_2}{dt} + \omega_0^2 r_2 = \xi r_1^2 \qquad (2.8)$$

Thus we find that the term ξr^2 in equation 2.4 causes a displacement that is nonlinear in E; this nonlinearity is $r_2 = a_2 E^2$. If we were to put terms of higher order in r into equation 2.4, we would find higher order nonlinearities also.

2.4 DEFINITION OF THE ELECTRIC FIELD

In the linear case, to avoid unnecessary algebra, we defined an electric field with a frequency ω and a complex amplitude, equation 1.2 Now, however, we want to examine the interaction of several fields with different frequencies, and so we have to define an electric field E with many components, each having its own phase and frequency. This gives

$$E = E(\omega_1)e^{-i\omega_1 t} + E^*(\omega_1)e^{i\omega_1 t} + E(\omega_2)e^{-i\omega_2 t}$$
$$+ E^*(\omega_2)e^{i\omega_2 t} + \cdots + E(\omega_n)e^{-i\omega_n t} + E^*(\omega_n)e^{i\omega_n t} \qquad (2.9)$$

and again, even with this complex notation, the algebra would become very complicated. To overcome this we define $-\omega_n = \omega_{-n}$, from which it follows that

$$E^*(\omega_n) = E(\omega_{-n}) \tag{2.10}$$

since $E^*(\omega_n) = E(-\omega_n)$. Using this definition, we can write equation 2.9 as a summation:

$$E = \sum_n E(\omega_n)e^{-i\omega_n t} \tag{2.11}$$

where the summation index runs over both the positive and negative values: $n = \pm 1, \pm 2, \pm 3$, etc. Now, using equation 2.10, the terms with negative indices neatly take care of the complex conjugate.

2.5   THE NONLINEAR POLARIZATION

Using equation 2.11 in equation 2.6, we find

$$\frac{dr_1}{dt} = -ia_1 \sum_n \omega_n E(\omega_n)e^{-i\omega_n t}$$

and

$$\frac{d^2 r_1}{dt^2} = -a_1 \sum_n \omega_n^2 E(\omega_n)e^{-i\omega_n t}$$

Therefore equation 2.7 gives

$$-a_1 \sum \omega_n^2 E(\omega_n)e^{-i\omega_n t} - 2a_1 i\gamma \sum \omega_n E(\omega_n)e^{-i\omega_n t} + a_1\omega_0^2 \sum E(\omega_n)e^{-i\omega_n t}$$

$$= -\frac{e}{m} \sum E(\omega_n)e^{-i\omega_n t} \tag{2.12}$$

However, since the sum of the individual terms at each frequency on the left-hand side of equation 2.12 has to be equal to the sum of the individual terms of the same frequency on the right-hand side, we can write

$$a_1 \sum E(\omega_n)e^{-i\omega_n t} = -\frac{e}{m} \sum \frac{E(\omega_n)e^{-i\omega_n t}}{\omega_0^2 - 2i\omega_n\gamma - \omega_n^2}$$

or

$$r_1 = -\frac{e}{m} \sum \frac{E(\omega_n)e^{-i\omega_n t}}{\omega_0^2 - 2i\omega_n\gamma - \omega_n^2} \tag{2.13}$$

Equation 2.13 is the result we found in Chapter 1 for the linear case, extended to more than one frequency. This result is now used to find the term of next higher order. To do this we substitute equation 2.13 in equation

2.8, which we then solve using the relation

$$\left(\sum_n E(\omega_n)e^{-i\omega_n t}\right)^2 = \sum_n \sum_m E(\omega_n)E(\omega_m)e^{-i(\omega_n+\omega_m)t} \tag{2.14}$$

where m has the same range of values as n.

From equations 2.13, 2.14, and 2.8, it follows that

$$r_2 = -\frac{e^2\xi}{m^2}\sum_n\sum_m \frac{E(\omega_n)E(\omega_m)e^{-i(\omega_n+\omega_m)t}}{F(\omega_0, \omega_n, \omega_m, \gamma)}$$

where

$$F(\omega_0, \omega_n, \omega_m, \gamma) = (\omega_0^2 - 2i\omega_n\gamma - \omega_n^2)(\omega_0^2 - 2i\omega_m - \omega_m^2)$$
$$\times [\omega_0^2 - 2i(\omega_n + \omega_m)\gamma - (\omega_n + \omega_m)^2]$$

For the polarization density we can write a power series also:

$$P = \sum_{l=1}^{\infty} P_e$$

with

$$P_e = -Ner_e$$

And so we find for the linear polarization

$$P_{\text{linear}} = \sum \chi^{(1)}(\omega_n)E(\omega_n)e^{-i\omega_n t}$$

where

$$\chi^{(1)}(\omega_n) = \frac{Ne^2}{m}\frac{1}{\omega_0^2 - 2i\gamma\omega_n - \omega_n^2}$$

and for the second-order polarization

$$P_{\text{second}} = \sum_n\sum_m \chi^{(2)}(\omega_n, \omega_m)E(\omega_n)E(\omega_m)e^{-i(\omega_n+\omega_m)t} \tag{2.15}$$

where

$$\chi^{(2)}(\omega_n, \omega_m) = -\frac{m\xi}{N^2e^3}[\chi^{(1)}(\omega_n)][\chi^{(1)}(\omega_m)][\chi^{(1)}(\omega_n + \omega_m)] \tag{2.16}$$

This second-order polarization is due to the nonlinear term ξr^2 in equation 2.4. Equation 2.15 shows that the second-order polarization density contains terms with all the frequencies $(\omega_n + \omega_m)$ possible for values of n and m of ± 1 and ± 2. Although in this book we are concerned with this second-order polarization only, higher orders do exist. If, for example, a third-order nonlinearity is added in equation 2.4, we find a third-order polarization term containing all the frequencies $\omega_n + \omega_m + \omega_p$ possible for different values of n, m, and p, and so on, for higher orders.

Examination of equation 2.16 shows that the second-order susceptibility $\chi^{(2)}$ depends on the product of the first-order susceptibilities for the three frequencies involved in the interaction. In Chapter 1, when the frequency dependence of the first-order susceptibility was discussed, we pointed out that, except for frequencies close to the resonant frequency ω_0, the linear susceptibility is purely real. Assuming that we operate with all three frequencies in lossless regions of the crystal, we see from equation 2.16 that the second-order susceptibility will also be purely real, $\chi^{(2)}(\omega_n, \omega_m) = \chi^{(2)}(-\omega_n, -\omega_m)$. Using this condition in equation 2.15 and setting m and n equal to $\pm 1, \pm 2$, we obtain the same result that was found in Section 2.2.

Henceforth, to simplify the notation, we use P for the second-order polarization and χ for the second-order susceptibility, unless otherwise indicated.

Equation 2.16 also allows a drastic reduction in the number of independent components of the nonlinear susceptibility. To be able to appreciate this reduction fully, we first extend our theory to three dimensions, and we also take into account the interaction of the generated wave with the incident waves.

2.6 EXTENSION TO THREE DIMENSIONS AND THREE MUTUALLY INTERACTING FIELDS

In the one-dimensional case, P, E, and χ were scalars. In three dimensions, P and E have a direction associated with them; they become vectors. Consequently χ, linking one vector to the product of two others, becomes a tensor of the third rank. So instead of equation 2.15, we write

$$P_i(\omega_{n+m}) = \sum_{jk} \sum_{nm} \chi_{ijk}(\omega_{n+m}, \omega_n, \omega_m)E_j(\omega_n)E_k(\omega_m)e^{-i(\omega_n+\omega_m)t} \quad (2.17)$$

where i, j, and k each take the values x, y, and z, and where $\omega_{n+m} = \omega_n + \omega_m$. Note that we have changed the notation from $\chi(\omega_n, \omega_m)$ as used in equation 2.16 to $\chi(\omega_{n+m}, \omega_n, \omega_m)$, where the dependence of χ on all three frequencies is clearly indicated.

Now if we include the interaction of $E(\omega_{n+m})$ with $E(\omega_n)$ and with $E(\omega_m)$, we find three interactions for each value of n and m; namely, $E(\omega_{n+m})$ generated by $E(\omega_n)$ and $E(\omega_m)$, $E(\omega_n)$ generated by $E(\omega_{m+n})$, and $E(\omega_m)$ and $E(\omega_m)$ generated by $E(\omega_{m+n})$ and $E(\omega_n)$. If we were to put all the different combinations into equation 2.17, we would obtain a great many frequency components of the nonlinear polarization. To limit the number of equations, we write here only the nonlinear polarization terms involved in the sum frequency generation $\omega_1 + \omega_2 = \omega_3$. These terms are

$$P_i(\omega_1) = \chi_{ijk}(\omega_1, -\omega_2, \omega_3)E_j(-\omega_2)E_k(\omega_3)e^{-i(\omega_3-\omega_2)t}$$
$$+ \chi_{ijk}(\omega_1, \omega_3, -\omega_2)E_j(\omega_3)E_k(-\omega_2)e^{-i(\omega_3-\omega_2)t}$$
$$P_j(\omega_2) = \chi_{jki}(\omega_2, \omega_3, -\omega_1)E_k(\omega_3)E_i(-\omega_1)e^{-i(\omega_3-\omega_1)t}$$
$$+ \chi_{jki}(\omega_2, -\omega_1, \omega_3)E_k(-\omega_1)E_i(\omega_3)e^{-i(\omega_3-\omega_1)t}$$
$$P_k(\omega_3) = \chi_{kij}(\omega_3, \omega_1, \omega_2)E_i(\omega_1)E_j(\omega_2)e^{-i(\omega_1+\omega_2)t}$$
$$+ \chi_{kij}(\omega_3, \omega_2, \omega_1)E_i(\omega_2)E_j(\omega_1)e^{-i(\omega_1+\omega_2)t} \quad (2.18)$$

and three more terms for the negative frequencies.

Note that the Einstein summation convention is used in equations 2.18. Thus $P_i(\omega_1)$, for example, is the sum of nine terms, using, respectively, the susceptibilities χ_{ixx}, χ_{ixy}, χ_{ixz}, χ_{iyx}, etc. Moreover, there are three different components of $P(\omega_1)$—that is, $P_y(\omega_1)$, $P_x(\omega_1)$, and $P_z(\omega_1)$—and the same is true for $P(\omega_2)$ and $P(\omega_3)$. Thus, even though we have assumed ω_1, ω_2, and ω_3 to be far enough away from any resonance frequency for $\chi_{ijk}(\omega_m, \omega_n)$ to be equal to $\chi_{ijk}(-\omega_n, -\omega_m)$, there could still be 81 different independent nonlinear susceptibilities in the three equations 2.18. Fortunately this is not so, since it follows from equation 2.16 and from the fact that the first-order susceptibility is real that

$$\chi_{ijk}(\omega_1, -\omega_2, \omega_3) = \chi_{jki}(\omega_2, \omega_3, -\omega_1) = \chi_{kij}(\omega_3, \omega_1, \omega_2) \quad (2.19)$$

In other words, the frequencies may be freely permuted, provided the cartesian indices i, j, and k are permuted with the frequencies.[5] This reduces the number of independent components to 27.

Equation 2.19 ties together a number of seemingly unrelated effects. To illustrate this, we consider the interaction of a wave with frequency ω_1 with itself. Apart from the second harmonic at $2\omega_1$ there will also be a direct polarization generated, equation 2.3. The nonlinear susceptibility for this "optical rectification" is $\chi_{ijk}(0, \omega_1, -\omega_1)$. Now from equation 2.19 we find

$$\chi_{ijk}(0, \omega_1, -\omega_1) = \chi_{jki}(\omega_1, -\omega_1, 0) \quad (2.20)$$

The right-hand side of equation 2.20 is the nonlinear susceptibility for an interaction where a direct field interacts with a wave with frequency ω_1 to produce a nonlinear polarization at ω_1. This nonlinear polarization generates a wave with frequency ω_1 but with a phase different from the phase of the input wave at ω_1. The end result is a phase change in the transmitted wave, or an apparent change in the index of refraction of the crystal. This effect will be recognized as the electrooptic effect (Section 1.11). Thus we see that the coefficients for optical rectification[12] are equal to those for the electrooptic effect, providing the indices are interchanged properly. (See also Section 2.17.)

2.7 MILLER'S RULE

From equation 2.16, and generalizing to three dimensions, we find

$$\chi_{ijk}^{(2)}(\omega_1, \omega_2, \omega_3) = [\chi_{ii}^{(1)}(\omega_1)][\chi_{jj}^{(1)}(\omega_2)][\chi_{kk}^{(1)}(\omega_3)]\Delta_{ijk} \quad (2.21)$$

This relation is known as Miller's rule, after R. C. Miller, who found empirically that the factor Δ_{ijk} is almost a constant for a wide range of materials.[61,116] This rule has been very important in the search for new nonlinear materials, since it states basically that materials with high refractive index also have a high nonlinear coefficient. However, this property is not as advantageous as it may seem (see Appendix 2).

2.8 THE COEFFICIENTS USED EXPERIMENTALLY

A great deal of confusion exists in the literature on the subject of the relation between the nonlinear susceptibility χ, used by theoreticians, and the
nonlinear coefficient d, which is used by experimentalists. Depending on the
particular definition of d used, a factor of $\frac{1}{2}$ does or does not enter the consideration, and since d^2 appears in the equation for the generated power,
this can lead to an error of a factor of 4. To avoid confusion here, we derive
the relation between d and χ as used in this book.

Taking the first one of equations 2.2, we can write

$$\mathscr{P}_{2\omega_1} = d\mathscr{E}_1{}^2 \cos\left[2(\omega_1 t + k_1 z)\right] = P(2\omega_1) + P^*(2\omega_1)$$

where

$$P(2\omega_1) = \tfrac{1}{2}d\mathscr{E}_1{}^2 e^{-2i(\omega_1 t + k_1 z)}$$

and from equation 2.18 we have

$$P(2\omega_1) = \chi(2\omega_1, \omega_1, \omega_1)E(\omega_1)E(\omega_1)e^{-2i\omega_1 t}$$

which, with our definition of $E(\omega)$ (equation 1.2), gives

$$P(2\omega_1) = \tfrac{1}{4}\chi(2\omega_1, \omega_1, \omega_1)\mathscr{E}^2 e^{-2i(\omega_1 t + k_1 z)}$$

Thus we find

$$d = \tfrac{1}{2}\chi(2\omega_1, \omega_1, \omega_1)$$

Providing the frequencies involved all lie within the "optical transmission
region" of the crystal, it is usually a very good approximation to leave aside
all frequency dependences of the susceptibility and to set the susceptibilities
equal, and so we write, generalizing to three dimensions,

$$d_{ijk} = \tfrac{1}{2}\chi_{ijk}$$

2.9 CONTRACTION OF THE INDICES

Consideration of equation 2.20 serves to emphasize the arbitrariness of the
sequence in which we have listed ω_1 and $-\omega_1$. And, in general, in an interaction between two fields $E(\omega_1)$ and $E(\omega_2)$ the sequence of these fields does
not give a physically discernible difference. We cannot state that one field
was applied first and then the other. Strictly speaking, then, we should modify
our equations in a way that brings out this point. This can be done by having
d operate on a column vector $\mathbf{F}$ which is a function of $\mathbf{E}(\omega_1)$ and $\mathbf{E}(\omega_2)$,
irrespective of their sequence:

$$F_\ell = (1 - \tfrac{1}{2}\delta_{jk})(E_j E_k + E_k E_j) \qquad (2.22)$$

where δ_{jk} is the Kronecker symbol: $\delta_{jk} = 1$ for $j = k$ and $\delta_{jk} = 0$ for $j \neq k$.

We can now set $d_{ijk} = d_{ikj}$, thus reducing the number of independent components to 18. As a result we can contract the last two indices using the scheme introduced in Section 1.10 for the contraction of the indices of the electrooptic coefficient—that is, by using the following values of l: $l = 1$ when $j = k = x$, $l = 2$ when $j = k = y$, $l = 3$ when $j = k = z$, $l = 4$ when $j = y$ and $k = z$, $l = 5$ when $j = x$ and $k = z$, $l = 6$ when $j = x$ and $k = y$.

Now, using equations 2.1, 2.22, and this contraction, we write

$$P_i = 2d_{i1}F_1 \tag{2.23}$$

The column vector $\mathbf{F}$ was introduced as a mathematical convenience to make sure that the sequence of the two interacting fields would not be important. This may seem like a theoretical nicety, but it does have practical importance as we show below.

In the case of second harmonic generation, we can write equations 2.22 and 2.23 in matrix form:

$$
\begin{vmatrix} \mathscr{P}_x \\ \\ \mathscr{P}_y \\ \\ \mathscr{P}_z \end{vmatrix}
=
\begin{vmatrix} d_{11} & d_{12} & d_{13} & d_{14} & d_{15} & d_{16} \\ \\ d_{21} & d_{22} & d_{23} & d_{24} & d_{25} & d_{26} \\ \\ d_{31} & d_{32} & d_{33} & d_{34} & d_{35} & d_{36} \end{vmatrix}
\begin{vmatrix} \mathscr{E}_x^2 \\ \mathscr{E}_y^2 \\ \mathscr{E}_z^2 \\ 2\mathscr{E}_y\mathscr{E}_z \\ 2\mathscr{E}_x\mathscr{E}_z \\ 2\mathscr{E}_x\mathscr{E}_y \end{vmatrix}
\tag{2.24}
$$

In the case of sum- or difference-frequency generation, however, this matrix notation is not strictly valid and should be used with care. If the sum frequency, for example, is generated between two waves $\mathscr{E}_x(\omega_1)$ polarized in the x direction and $\mathscr{E}_y(\omega_2)$ polarized in the y direction, then

$$\mathscr{P}_1(\omega_1 + \omega_2) = d_{16}\mathscr{E}_x(\omega_1)\mathscr{E}_y(\omega_2) \tag{2.25}$$

The factor 2 does not appear here because

$$\mathscr{E}_y(\omega_1)\mathscr{E}_x(\omega_2) = 0 \tag{2.26}$$

Thus caution should be exercised whenever the two input frequencies are physically separable.

2.10 CRYSTAL SYMMETRY

Summarizing our conclusions from the previous sections, we find that we have a maximum number of 18 independent tensor elements, and that crystals which have a center of symmetry cannot exhibit a second-order polarization. Of the 32 different crystal classes, 21 are noncentrosymmetrical, but of these only one has no symmetry at all; this is class 1 in the triclinic system. For all the other classes there are one or more symmetry operations which transform the crystal into itself. Obviously, if a crystal class has a given susceptibility matrix, and we perform a symmetry operation that does not physically alter the crystal, then the same operation should also leave the matrix unchanged. As a result of this, certain components of the matrix must be zero, and others are equal or numerically equal but opposite in sign. By performing the allowed symmetry operations for each crystal class,[89] we find a matrix of a given form for each one of the 21 noncentrosymmetric crystal classes. Alpha-iodic acid, for example, belongs to the crystal class 222. This class has 3 diad axes, parallel to the x, the y, and the z axes, respectively. This means that if we take an iodic acid crystal and rotate it 180° around any one of these three axes, the crystal will appear exactly the same as it did before the rotation. Thus, if before a 180° rotation around the z axis we had a field E_x in the x direction, the same field will be applied in the $-x$ direction after the rotation. As a result, $d_{111}E_x^2$ has to be equal to $-d_{111}E_x^2$, and we know that this can only be so if $d_{111} = 0$. Similarly, we find from this rotation that d_{112}, d_{122}, d_{133}, d_{211}, d_{212}, d_{222}, d_{233}, d_{313}, and d_{323} all have to be zero. Thus a diad axis parallel to the z axis leaves only those tensor elements which have either one or three 3's as subscripts such as d_{333} and d_{223}. Similarly, a diad axis parallel to the x axis only leaves tensor elements that have either one or three 1's.

Taking all three diad axes in crystal class 222 into account, we find that only three tensor elements are nonzero; namely, d_{123}, d_{213}, and d_{312} (or, in contracted notation, d_{14}, d_{25}, and d_{36}).

An even more sweeping symmetry condition applies to nonlinear processes in which the nonlinear polarization is purely electronic in origin and in which the crystal is lossless throughout a spectral region that includes all the frequencies involved in the interaction. This condition, first formulated by Kleinman,[90] states that in such cases the elements of the tensor d_{ijk} formed by freely permuting i, j, and k are all equal. This is not the same as equation 2.19, since here only the cartesian indices and not the frequencies are permuted; however, comparison with equation 2.16 shows that Kleinman's conjecture is equivalent to stating that the dispersion of the nonlinearity is negligible. It has proven to be correct to within measurement accuracy. In only one case has a deviation been measured, but this may have been due to a dispersion in the nonlinear susceptibility.[131] If an absorption band lies in between the frequencies involved in the interaction, Kleinman's symmetry condition naturally does not hold. Thus it does not apply to the electrooptic effect or to difference-frequency generation of far-infrared wavelengths.[173]

For a more thorough treatment of these symmetry requirements, the reader is referred to the book by Nye.[130]

Equation 2.24 suggests that the nonlinear susceptibility matrix has the same form as the piezoelectric matrix, and indeed, for a given crystal class, the matrix for the second-order susceptibility is homologous with the piezoelectric matrix except for the factors of 2 in the column vector for the field. In the piezoelectric case these factors of 2 are included in the matrix itself. Thus we can find the independent elements of the nonlinear matrix by looking up the piezoelectric matrix and omitting the factors of 2. For class $3m$, for example, the piezoelectric matrix yields[130] $d_{11} = -d_{12} = -2d_{26}$; but in the nonlinear matrix this becomes $d_{11} = -d_{12} = -d_{26}$.

Note that the numerical values of the matrix elements for the two effects are not related at all.

The nonzero elements of the tensor d_{ij} for all the crystal classes are listed in Figure 2.1, both for the cases in which the Kleinman symmetry condition obtains and for those in which it does not.

3.11 DEFINITION OF d_{eff}

In the foregoing discussion we have successfully restricted the initially large number of components to 18 in the most general case and to far fewer in most practical cases. However, the expression for the polarization, equation 2.17, still involves a summation to account for the nonzero matrix elements and the directions of polarization of the interacting waves. We can

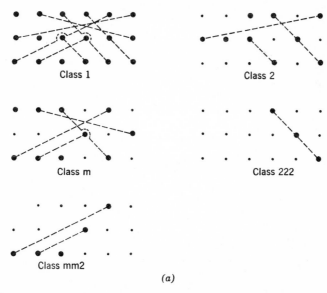

Class 1

Class 2

Class m

Class 222

Class mm2

(a)

Figure 2.1a

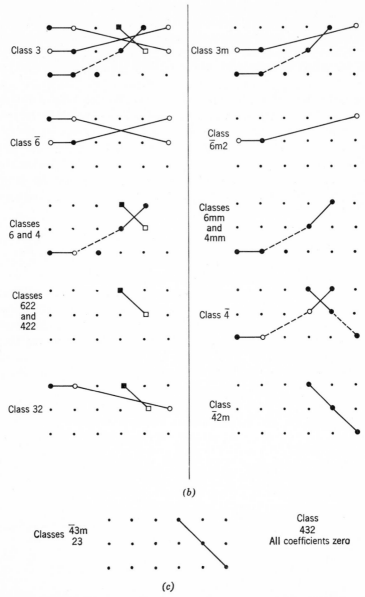

Figure 2.1 Form of the d_{ij} matrix for the different crystal classes: (a) Biaxial crystal classes, (b) uniaxial crystal classes (c) isotropic crystal classes. Legend: small dot = zero coefficient; square dot = coefficient that is zero when Kleinman's symmetry condition is valid; connected dots = numerically equal coefficients, but the open-dot coefficient is opposite in sign to the closed dot to which it is joined. Dashed connections are valid only under Kleinman's symmetry conditions. Since in all biaxial classes the coefficients are all independent when Kleinman's symmetry is not valid, the dashed lines in class 1 show the influence of Kleinman's symmetry on the 18 coefficients.

now simplify the notation even further by writing a new expression:

$$\mathscr{P} = d_{\text{eff}}\mathscr{E}(\omega_1)\mathscr{E}(\omega_2) \tag{2.27}$$

in which all these summations have been performed for the particular case at hand.*

The advantage of this notation is that it reduces the problem to one dimension. All further derivations are made using this effective nonlinearity, yielding in the end a simple, universally valid expression for the generated power. For given experimental conditions, the appropriate equation for d_{eff} is substituted in this equation, restoring again the full three-dimensional aspect of the problem. A table of equations for d_{eff} to be used in the various possible cases is given in Chapter 3.

2.12 AN EXAMPLE

As an example of some of the points mentioned previously, we now present an actual case.

Suppose we have two linearly polarized plane waves with amplitudes $\mathscr{E}'$ and $\mathscr{E}''$ propagating in a quartz crystal at an angle θ to the optic axis. Suppose moreover that the plane through the optic axis and the incident rays makes an angle φ with the x axis and that $\mathscr{E}'$ lies in this plane (e-ray) and $\mathscr{E}''$ is perpendicular to this plane (o-ray). We want to know the nonlinear polarization at the sum frequency due to the interaction of these two fields.

From Figure 2.1 or from piezoelectric data we know the nonlinear susceptibility matrix for quartz (class 32)

$$\begin{vmatrix} d_{11} & -d_{11} & 0 & d_{14} & 0 & 0 \\ 0 & 0 & 0 & 0 & -d_{14} & -d_{11} \\ 0 & 0 & 0 & 0 & 0 & 0 \end{vmatrix}$$

Writing out the three components of the nonlinear polarization, using the noncontracted form of the coefficient, we have

$$\mathscr{P}_1 = d_{111}\mathscr{E}'_1\mathscr{E}''_1 + d_{122}\mathscr{E}'_2\mathscr{E}''_2 + d_{123}\mathscr{E}'_2\mathscr{E}''_3 + d_{132}\mathscr{E}'_3\mathscr{E}''_2$$

$$\mathscr{P}_2 = d_{213}\mathscr{E}'_1\mathscr{E}''_3 + d_{231}\mathscr{E}'_3\mathscr{E}''_1 + d_{212}\mathscr{E}'_1\mathscr{E}''_2 + d_{221}\mathscr{E}'_2\mathscr{E}''_1$$

$$\mathscr{P}_3 = 0 \tag{2.28}$$

---

* Since $\mathscr{P}$ is the polarization for only one frequency component and not the total nonlinear polarization, the factor 2 of equation 2.23 does not appear here. See also equation 2.2.

where

$$d_{111} = -d_{122}$$
$$d_{123} = -d_{213}$$
$$d_{212} = d_{221} = -d_{111}.$$

The different components of the fields are

$$\mathscr{E}'_1 = \mathscr{E}' \cos \theta \cos \varphi$$
$$\mathscr{E}'_2 = \mathscr{E}' \cos \theta \sin \varphi$$
$$\mathscr{E}'_3 = -\mathscr{E}' \sin \theta$$
$$\mathscr{E}''_1 = -\mathscr{E}'' \sin \varphi$$
$$\mathscr{E}''_2 = \mathscr{E}'' \cos \varphi$$
$$\mathscr{E}''_3 = 0 \tag{2.29}$$

Substitution of equation 2.29 in equation 2.28 gives

$$\mathscr{P}_1 = \mathscr{E}'\mathscr{E}''(-d_{11} \cos \theta \sin 2\varphi - d_{14} \sin \theta \cos \varphi)$$
$$\mathscr{P}_2 = \mathscr{E}'\mathscr{E}''(-d_{11} \cos \theta \cos 2\varphi - d_{14} \sin \theta \sin \varphi)$$
$$\mathscr{P}_3 = 0$$

Note that, since $\mathscr{E}''_3 = 0$, the two terms containing $\sin \varphi$ have d_{14} and not $2d_{14}$, as might have been deduced from the contracted matrix. This emphasizes the point made in equations 2.25 and 2.26.

Now, since

$$\mathscr{P}_\perp = \mathscr{P}_1 \sin \varphi - \mathscr{P}_2 \cos \varphi$$
$$\mathscr{P}_\parallel = (\mathscr{P}_1 \cos \varphi + \mathscr{P}_2 \sin \varphi) \cos \theta + \mathscr{P}_3 \sin \theta$$

where $P_\parallel$ is the nonlinear polarization parallel to $\mathscr{E}'$, we have

$$\mathscr{P}_\parallel = \mathscr{E}'\mathscr{E}''(d_{11} \cos \theta \sin 3\varphi - d_{14} \sin \varphi \cos \theta) \tag{2.30}$$
$$\mathscr{P}_\perp = \mathscr{E}'\mathscr{E}''(d_{11} \cos \theta \cos 3\varphi) \tag{2.31}$$

The correctness of these two equations can be ascertained by direct examination. In quartz the z axis is an axis of threefold symmetry. This is borne out by the appearance of the argument 3φ, which shows that $\mathscr{P}_\perp$, for example, is the same for $\varphi = 0°$, $120°$, $240°$, and so on.

Equations 2.30 and 2.31 quite clearly reveal the complicated form of the nonlinearity in an actual case. In our notation, using the effective nonlinearity, we write $\mathscr{P} = d_{\text{eff}}\mathscr{E}\mathscr{E}$ for equations 2.30 and 2.31 and then, at the end of the calculation, we substitute the correct value for d_{eff}. In this case the choice would depend on whether $\mathscr{P}_\parallel$ or $\mathscr{P}_\perp$ is required.

Equations 2.30 and 2.31 could also have been arrived at by rotating the tensor first over an angle φ around the z axis and then over an angle θ around the new y axis. The two fields then lie in the new x and z directions, and so the polarizations are found by a simple multiplication of the rotated tensor by $\mathscr{E}_1$ and $\mathscr{E}_3$. Since we are only interested in $\mathscr{P}_1$ and $\mathscr{P}_3$ in the new coordinate system, only six terms of the rotated tensor have to be calculated: d_{11}, d_{13} d_{15}, and d_{31}, d_{33}, d_{35}. Even so, this method is usually more time consuming than the one used previously.

Henceforth, unless otherwise indicated, we use d for the effective non-linearity.

2.13 THE COUPLED AMPLITUDE EQUATIONS

Thus prepared with all possible simplifications, we are ready to derive equations for the electromagnetic radiation generated by the nonlinear polarization. Clearly, at any point, the decrease or increase of the amplitude of a wave at a particular frequency depends on the amplitudes of two other waves. For three interacting waves, then, we must seek three coupled amplitude equations, each giving the rate of growth, or decay, of the field at one frequency as a function of the fields at the two other frequencies. In addition, we should expect in each of these equations some measure of the phase difference between the polarization wave and the electromagnetic wave.

The method followed is very much the same as in the case of the linear refractive index, discussed in Chapter 1. We start by introducing the non-linear polarization as a source term in Maxwell's equations

$$\mathbf{\nabla} \times \mathbf{H} = \frac{1}{c} \frac{\partial \mathbf{D}}{\partial t} \tag{2.32}$$

$$\mathbf{\nabla} \times \mathbf{E} = -\frac{1}{c} \frac{\partial}{\partial t} (\mu \mathbf{H}) \tag{2.33}$$

$$\mathbf{D} = \varepsilon \mathbf{E} + 4\pi \mathbf{P}$$

Here the linear polarization is included in ε and $\mathbf{P}$ is only the nonlinear polarization.

Again, as we did in Chapter 1, we assume that the material is noncon-ducting, and we take the curl of both sides of equation 2.33. Thus for non-magnetic materials we find

$$\nabla^2 \mathbf{E} = \frac{\partial^2}{\partial t^2} \left(\frac{\varepsilon}{c^2} \mathbf{E} \right) - \frac{4\pi}{c^2} \frac{\partial^2 \mathbf{P}}{\partial t^2} \tag{2.34}$$

We now restrict the problem to one dimension by taking $\partial/\partial y = \partial/\partial x = 0$ and assuming propagation in the z direction. Moreover, to limit the discussion to three interacting traveling waves, we define

$$E_1(z, t) = E_1(z)e^{-i(\omega_1 t - k_1 z)}$$
$$E_2(z, t) = E_2(z)e^{-i(\omega_2 t - k_2 z)}$$
$$E_3(z, t) = E_3(z)e^{-i(\omega_3 t - k_3 z)} \tag{2.35}$$

where the subscripts 1, 2, and 3 refer to the frequencies. There are, of course, three similar equations for the negative frequencies.

This definition is very similar to the one used in Chapter 1, except that in a linear medium the complex amplitude is constant whereas here the complex amplitude changes as a result of the interaction with waves at different frequencies. Quite apart from the phase determined by the propagation constant k, the wave has a phase φ which is dependent on z. We have

$$E_1(z) = \tfrac{1}{2}\mathscr{E}_1(z)e^{i\varphi_1(z)}$$

From equation 2.18 we find

$$P_1(z, t) = 4d\, E_2^*(z)E_3(z)e^{-i[(\omega_3 - \omega_2)t - (k_3 - k_2)z]}$$
$$P_2(z, t) = 4d\, E_3(z)E_1^*(z)e^{-i[(\omega_3 - \omega_1)t - (k_3 - k_1)z]}$$
$$P_3(z, t) = 4d\, E_1(z)E_2(z)e^{-i[(\omega_1 + \omega_2)t - (k_1 + k_2)z]} \tag{2.36}$$

where again we have $\omega_1 + \omega_2 = \omega_3$.

From equations 2.36 we find

$$\frac{\partial^2 P_1}{\partial t^2} = -(\omega_3 - \omega_2)^2 4d\, E_2^*(z)E_3(z)e^{-i[(\omega_3 - \omega_2)t - (k_3 - k_2)z]} \tag{2.37}$$

and similar equations for $\partial^2 P_2/\partial t^2$ and $\partial^2 P_3/\partial t^2$.

Assuming that the variation of the complex field amplitude with z is small enough so that $k\, dE/dz \gg d^2E/dz^2$ we find

$$\frac{\partial^2 E_1(z, t)}{\partial z^2} = -\left[k_1^2 E_1(z) - 2ik_1 \frac{dE_1(z)}{dz} \right] e^{-i(\omega_1 t - k_1 z)} \tag{2.38}$$

Recognizing that $\varepsilon\omega_1^2/c^2 = k_1^2$, and assuming that equation 2.34 is satisfied by each frequency component separately, we find from equations 2.34, 2.37, and 2.38

$$\frac{dE_1(z)}{dz} = -i\frac{8\pi\omega_1^2}{k_1 c^2} d\, E_2^*(z)E_3(z)e^{i(k_3 - k_2 - k_1)z}$$

$$\frac{dE_2(z)}{dz} = -i\frac{8\pi\omega_2^2}{k_2 c^2} d\, E_1^*(z)E_3(z)e^{i(k_3 - k_2 - k_1)z}$$

$$\frac{dE_3(z)}{dz} = -i\frac{8\pi\omega_3^2}{k_3 c^2} d\, E_1(z)E_2(z)e^{i(k_1 + k_2 - k_3)z} \tag{2.39}$$

These are the three coupled amplitude equations sought. We see that each equation does indeed give the rate of change with distance of the amplitude at one frequency as a function of the amplitudes at the two other frequencies and of the phase difference between the polarization wave and the electromagnetic wave. We write

$$\Delta k = k_3 - k_2 - k_1 \tag{2.40}$$

Clearly all three amplitudes are dependent on one another. As a result, a general solution of these three equations is anything but simple. However, if we assume that the amount of power generated is small enough so that the amplitudes of the two input frequencies are essentially constant, then the three equations reduce to one, which can be integrated quite easily.

Let us take, for example, the interaction $\omega_1 + \omega_2 = \omega_3$. Since we asume that E_1 and E_2 are constant, we can write

$$E_3 = -\frac{8\pi i \omega_3^2}{k_3 c^2} d\, E_1 E_2 \int_0^L e^{i\,\Delta kz}\, dz \tag{2.41}$$

which can easily be integrated to give

$$E_3 = -\frac{8\pi \omega_3^2}{k_3 c^2\, \Delta k} d\, E_1 E_2 (e^{i\,\Delta kL} - 1)$$

where L is the length of the crystal. Since $\omega_3 = 2\pi c/\lambda_3$ and $k_3 = 2\pi n/\lambda_3$, we can also write

$$E_3 = -\frac{16\pi^2}{n_3 \lambda_3\, \Delta k} d\, E_1 E_2 (e^{i\,\Delta kL} - 1) \tag{2.42}$$

The power per unit area in a material with index n is

$$S = \frac{cn}{8\pi}\, \mathscr{E}^2 = \frac{cn}{2\pi}\, EE^* \tag{2.43}$$

Multiplying equation 2.42 by its complex conjugate and using equation 2.43 on both sides, we find for the output power per unit area

$$S_3 = \frac{512\pi^5 L^2 d^2 S_1 S_2}{n_1 n_2 n_3 \lambda_3^2 c}\left(\frac{\sin x}{x}\right)^2 \tag{2.44}$$

or if W is the total power and A is the area

$$W_3 = \frac{512\pi^5 L^2 d^2 W_1 W_2}{n_1 n_2 n_3 \lambda_3^2 cA}\left(\frac{\sin x}{x}\right)^2 \tag{2.45}$$

In both equations 2.43 and 2.44 we have $x = \Delta kL/2$.

It is important to obtain the correct dimensions in these equations. Since we have worked in the c.g.s. system, the total power W is given in ergs, d is in c.g.s., and all lengths are in centimeters.

Often it is convenient to give the powers in watts, and so we give here an equation with hybrid dimensions

$$P_3 = \frac{52.2d^2L^2P_1P_2}{n_1n_2n_3\lambda_3{}^2}\left(\frac{\sin x}{x}\right)^2 \tag{2.46}$$

where L and λ are in centimeters and d is in c.g.s., but P_1, P_2, and P_3, are in watts per square centimeter.

Equation 2.44 shows some features that are typical for bulk frequency mixing:

1. For $\Delta k \neq 0$ the output power varies as $(\sin x/x)^2$. This point is treated in detail in the next chapter.

2. For $k = 0$ the output is proportional to the square of the length of the crystal, as measured in wavelengths of output.

3. The output power is proportional to the product of the input powers.

Point 2, of course, is true only in the small-signal approximation that is used here.

Since the output signal from a frequency mixing experiment quite often is small compared with the inputs, it can at times be very difficult to ascertain whether an observed signal is indeed the derived output or is merely a scattered fraction of the input. Here point 3 provides a positive check: we observe the decrease in power of the observed signal for a reduction in power of the inputs. For example, if both inputs are decreased by a factor of 2, the output should go down by a factor of 4.

A further check on the mixed frequency nature of an observed output is provided by the factor d in equation 2.44, which gives, implicitly, the intensity and the direction of polarization of the output as a function of the polarizations and the directions of propagation of the inputs. This point is treated in detail in Chapter 3.

2.14   THE MANLEY–ROWE RELATIONS

Further examination of the three coupled equations (2.39) shows that the second equation can be obtained from the first by interchanging E_1 and E_2, but the third equation cannot be obtained by a similar interchange. The significance of this statement becomes very clear when we look at the power flow in the interaction. Assuming $\Delta k = 0$, we find from equations 2.39

$$\frac{n_1 c}{\omega_1} E_1^* \frac{dE_1}{dz} = -8\pi i\, d\, E_1^* E_2^* E_3 \tag{2.47}$$

$$\frac{n_2 c}{\omega_2} E_2^* \frac{dE_2}{dz} = -8\pi i\, d\, E_2^* E_1^* E_3 \tag{2.48}$$

$$\frac{n_3 c}{\omega_3} E_3^* \frac{dE_3}{dz} = -8\pi i\, d\, E_3^* E_2 E_1 \tag{2.49}$$

and since the right-hand sides of equations 2.47 and 2.48 are equal to the complex conjugate of the right-hand side of equation 2.49, we have

$$\frac{n_1 c}{\omega_1} \frac{d}{dz}(E_1 E_1^*) = \frac{n_2 c}{\omega_2} \frac{d}{dz}(E_2 E_2^*) = -\frac{n_3 c}{\omega_3} \frac{d}{dz}(E_3 E_3^*) \tag{2.50}$$

Now, using equation 2.43, we find

$$\frac{\text{change in power at } \omega_1}{\omega_1} = \frac{\text{change in power at } \omega_2}{\omega_2}$$

$$= -\frac{\text{change in power at } \omega_3}{\omega_3}$$

This relation, first formulated by Manley and Rowe,[106] has far-reaching consequences. Note that we have derived it from the coupled amplitude equations without specifying the particular interaction, and so the relation is valid for sum-frequency or difference-frequency generation. For sum-frequency generation (e.g., between two lasers with frequencies ω_1 and ω_2), the Manley–Rowe relation states that both lasers will lose power, which is gained by the sum frequency $\omega_3 = \omega_1 + \omega_2$. But for difference-frequency generation, $\omega_3 - \omega_2 = \omega_1$, we find from the same relation that the source at ω_3 loses power not only to the generated frequency ω_1, *but also to the source of ω_2.* In other words, if the difference frequency ω_1 is generated using two frequencies ω_3 and ω_2 as inputs, then *both ω_1 and ω_2 gain in power.*

Since $[EE^*/\omega]$ is a measure of the photon density, we can also say that the photon at ω_3 is split into a photon at ω_1 and a photon at ω_2, or, in the case of sum-frequency generation, that two photons—one at ω_1 and one at ω_2—combine to give one photon at ω_3.

Since the source of ω_2 in the difference-frequency generation gains power, this opens the possibility of generating the difference frequency between a strong source at ω_3, the pump, and a very weak source at ω_2. If the weak signal at ω_2 is passed through the nonlinear crystal again and again, it will gain power with every pass and will build up; as a consequence, the signal at ω_1 will build up also. Now the weak signal at ω_2 does not have to be brought in from the outside, it can be a frequency in the noise, and the many passes can be obtained by putting the crystal in a mirror cavity that is resonant at ω_2. If the gain per pass is higher than the loss per pass, such a system will oscillate. This is the *parametric oscillator*, described in detail in Chapter 7.

2.15 SECOND HARMONIC GENERATION

A special case of frequency mixing occurs when both input frequencies are equal. The frequency of the output is then twice the frequency of the input, and the interaction is known as second harmonic generation. To obtain the coupled amplitude equations for this special case it does not suffice to set $\omega_1 = \omega_2$ in equations 2.39, since this would give a polarization at 2ω, which would be too large by a factor of 2. This is because the sum frequency is made up of a term $\omega_1 + \omega_2$ and a term $\omega_2 + \omega_1$, whereas the second harmonic has only a term $\omega_1 + \omega_1$ taken once. Going back to equation 2.17 for the nonlinear polarization, and using the same steps as before, we obtain:

$$\frac{dE_1(z)}{dz} = -i\frac{8\pi\omega_1^2}{k_1 c^2} d\, E_1^*(z)E_2(z)e^{-i\,\Delta kz}$$

$$\frac{dE_2(z)}{dz} = -i\frac{16\pi\omega_1^2}{k_2 c^2} d\, E_1^2(z)e^{i\,\Delta kz} \tag{2.51}$$

where $\omega_2 = 2\omega_1$ and Δk is $2k_1 - k_2$.

Thus we find for the second harmonic power per unit area, in the small-signal approximation, the following relation:

$$S(2\omega) = \frac{512\pi^5 d^2 L^2 S^2(\omega)}{n(2\omega)n^2(\omega)\lambda^2 c}\left(\frac{\sin x}{x}\right)^2 \tag{2.52}$$

where λ is the wavelength of the fundamental and again $x = \Delta kL/2$. The fundamental frequency is ω.

The two coupled amplitude equations (2.51) can also be solved without making the small-signal approximation. By writing the real and the imaginary parts of the two equations, we find

$$\frac{d\mathscr{E}_1}{dz} = -\frac{4\pi\omega^2 d}{c^2 k_1}\,\mathscr{E}_1\mathscr{E}_2 \sin\theta \tag{2.53}$$

$$\frac{d\mathscr{E}_2}{dz} = \frac{8\pi\omega^2 d}{c^2 k_1}\,\mathscr{E}_1^2 \sin\theta \tag{2.54}$$

$$\frac{d\theta}{dz} = \Delta k - \frac{8\pi\omega^2 d}{c^2}\left[\frac{\mathscr{E}_2}{k_1} - \frac{\mathscr{E}_1^2}{\mathscr{E}_2 k_2}\right]\cos\theta \tag{2.55}$$

where $\theta = \Delta kz + 2\varphi_1(z) - \varphi_2(z)$ and $\Delta k = 2k_1 - k_2$. From equation 2.51 we can derive the Manley–Rowe relation using the method of Section 2.14. From it we find for the power flow

$$W = \frac{k_1 c^2}{8\pi\omega_1^2} \mathscr{E}_1^2(z) + \frac{k_2 c^2}{16\pi\omega_1^2} \mathscr{E}_2^2(z) \tag{2.56}$$

Substitution of equations 2.53 and 2.54 into equation 2.55 gives

$$\frac{d\theta}{dz} = \Delta k + \left(\frac{\cos\theta}{\sin\theta}\right)\frac{d}{dz}[\ln(\mathscr{E}_1^2\mathscr{E}_2)] \tag{2.57}$$

The solution of these equations was first worked out by Bloembergen and his co-workers;[5] the reader is referred to the original paper for the general solution. However, normally the interaction is phase matched and the initial second-harmonic power is zero. This makes for a much simpler solution, which we give here. Accordingly, we set $\Delta k = 0$, and $\mathscr{E}_2(0) = 0$, and use the substitutions

$$u = \frac{\mathscr{E}_1(z)}{\mathscr{E}_1(0)} \tag{2.58}$$

$$v = \frac{\mathscr{E}_2(z)}{\mathscr{E}_1(0)} \tag{2.59}$$

and

$$\zeta = \frac{4\pi\omega^2 d}{c^2 k_1} \mathscr{E}_1(0)z \tag{2.60}$$

From equation 2.56 we see that, since $k_2 = 2k_1$

$$u^2 + v^2 = 1 \tag{2.61}$$

Substituting equations 2.58 and 2.59 in equations 2.53 and 2.54, we can use equation 2.60 to find

$$\frac{du}{d\zeta} = -uv\sin\theta \tag{2.62}$$

$$\frac{dv}{d\zeta} = u^2\sin\theta \tag{2.63}$$

and equation 2.57 gives

$$\frac{d\theta}{d\zeta} = \frac{\cos\theta}{\sin\theta}\frac{d}{d\zeta}\ln(u^2 v) \tag{2.64}$$

Equation 2.63 can be integrated to give

$$-\ln(\cos\theta) = \ln(u^2 v) + C \tag{2.65}$$

where C is an integration constant. Writing equation 2.64 as $\Gamma = u^2v \cos \theta$, we see that the initial condition $v(0) = 0$ implies that $\Gamma = 0$. Using this and equation 2.61 in equation 2.63, we find

$$\frac{dv}{d\zeta} = \pm(1 - v^2)$$

which can be integrated to give

$$v = \tanh (\zeta + \zeta_0) \tag{2.66}$$

and

$$u = \text{sech} (\zeta + \zeta_0) \tag{2.67}$$

Since $v(0) = 0$ we have $\zeta_0 = 0$. Then using equations 2.58, 2.59, and 2.60, we find from equations 2.66 and 2.67 that

$$\mathscr{E}_2(L) = \mathscr{E}_1(0) \tanh \frac{L}{l_{\text{SH}}} \tag{2.68}$$

and

$$\mathscr{E}_1(L) = \mathscr{E}_1(0) \text{ sech} \frac{L}{l_{\text{SH}}} \tag{2.69}$$

where

$$l_{\text{SH}} = \left[\frac{4\pi\omega^2 d}{k_1 c^2} \mathscr{E}_1(0) \right]^{-1} \tag{2.70}$$

2.16 OUTPUT ANGLE

Thus far we have only shown that the nonlinear polarization generates an output at a frequency different from that of the input, but we have not considered the angle at which this output is radiated. For the linear refractive index, we learned in Chapter 1 that the oscillating dipoles in the Lorentz model form a phased array of antennas. In this chapter we added a nonlinear term to the Lorentz model and found that this resulted in frequency components which are higher harmonics, and mixtures of the input frequencies. Thus we now have an array of antennas that is still correctly phased for the input frequencies but also radiates other frequencies, for which there is in general a phase mismatch. This phase mismatch was the cause of the term Δk in equation 2.41.

In the next chapter we discuss ways of obtaining a phased array, $\Delta k = 0$, for at least one of these "nonlinear" frequency components. However, since the wave generated by this phased array has a frequency completely different from the frequencies of the input waves (and therefore a wavelength that is completely different from the wavelength of the input waves), it is of interest to determine the angle over which the output will be radiated. When the output frequency is of the same order of magnitude as the input frequencies, such as in second-harmonic generation, the difference in the wavelength is not too important, but in difference frequency generation, where the output may have a wavelength much larger than the wavelengths of the inputs, the questions is anything but trivial. It then becomes possible to focus the two input beams to a spot with a size smaller than the wavelength of the output. Thus we might expect the angle over which the output would be radiated to be much larger than the convergence angle of the input beams.

To find an expression for the output angle, we examine the phase-matched interaction between two waves with frequencies ω_2 and ω_3, generating a difference frequency $\omega_1 = \omega_3 - \omega_2$. To simplify the problem, we assume that the interaction takes place in a cylinder with radius a and length L and that in this cylinder the two waves at ω_2 and ω_3 are plane waves, parallel to each other, and having uniform intensities out to the edge. Obviously, then, the polarization wave at ω_1 is also a plane wave with uniform intensity out to the edge. We now derive the angular dependence of the radiation at ω_1 in a manner analogous to the calculation of the radiation pattern of a phased array of antennas.

We define a cylindrical coordinate system with the z axis normal to the wavefronts; the origin is centered at the input end of the interaction cylinder (Figure 2.2).

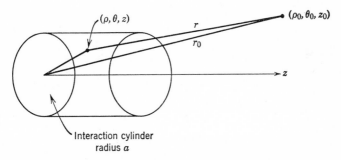

Figure 2.2 Coordinate system for the derivation of the output angle.

We now determine the field at an arbitrary point in space by summing the contributions to this field from every single point in the interaction cylinder. For this summation we need to know the amplitude of each contribution and its phase. If we set the phase of the wave equal to zero at $z = 0$, we find that the phase of a wave starting at a point $[\rho, \varphi, z]$ is given by $e^{-ik_1 z}$ at its starting point, and by $e^{-ik_1[z+r]}$ at an arbitrary point $[\rho_0, \theta_0, z_0]$, where r is the distance between $[\rho_0, \theta_0, z_0]$ and $[\rho, \theta, z]$. To find the field at $[\rho_0, \theta_0, z_0]$ due to the entire cylinder, we evaluate the integral

$$E = E_0 \int_V e^{-ik_1(z+r)} \, dV \tag{2.71}$$

where V is the volume of the cylinder.

Let r_0 be the distance from $[\rho_0, \theta_0, z_0]$ to the origin. Then

$$r_0{}^2 = \rho_0{}^2 + z_0{}^2$$

We now write

$$r = r_0 + \Delta r$$

$$z_0 = r_0 \cos \varphi$$

$$\theta_0 - \theta = \theta' \tag{2.72}$$

By applying the cosine law to the triangles formed by the points $[\rho, \theta, z]$, $[\rho_0, \theta_0, z_0]$, and the origin, we find

$$r^2 = r_0{}^2 + \rho^2 + z^2 - 2r_0(\rho \sin \varphi \cos \theta' + z \cos \varphi) \tag{2.73}$$

Combining equations 2.72 and 2.73 we arrive at a quadratic equation in Δr with the two roots

$$\Delta r = r_0 \pm r_0 \left(1 + \frac{\rho^2}{r_0^2} + \frac{z^2}{r_0^2} - \frac{2c}{r_0}\right)^{1/2} \tag{2.74}$$

Applying the binomial theorem to equation 2.74 and neglecting powers higher than the first of $1/r_0$, we find

$$\Delta r = -(\rho \sin \varphi \cos \theta' + z \cos \varphi) + \frac{[(\rho \sin \varphi \cos \theta' + z \cos \rho)^2 + \rho^2 + z^2]}{2r_0}$$

If the point $[\rho_0, \theta_0, z_0]$ is far enough away, we have the relation

$$\frac{k_1[(\rho \sin \varphi \cos \theta' + z \cos \varphi)^2 + \rho^2 + z^2]}{2r_0} \ll 2\pi$$

and we can write equation 2.71 as

$$E = E_0 e^{-ik_1 r_0} \int_V e^{-ik_1[(1 - \cos \varphi)z - \rho \sin \varphi \cos \theta']} \, dV \tag{2.75}$$

Integration of equation 2.75 and multiplication of the integral by its complex conjugate gives

$$EE^* = E_0{}^2 V^2 \left[\frac{2J_1(k_1 a \sin \varphi)}{k_1 a \sin \varphi}\right]^2 \left[\frac{\sin \{k_1 L(1 - \cos \varphi)/2\}}{k_1 L(1 - \cos \varphi)/2}\right] \tag{2.76}$$

where J_1 is a first-order Bessel function.

The term $[2J_1(k_1a \sin \varphi)]/k_1a \sin \varphi]^2$ is the same as the term that gives the Fraunhofer diffraction of a circular aperture with radius a.[82] The second bracketed term in the right-hand side of equation 2.76 is similar to the $[(\sin x)/x]^2$ term in equation 2.44, but Δk is replaced by $k_1 [1 - \cos \varphi]$. Thus we see that the angular distribution of the output radiation is determined by the Fraunhofer diffraction pattern of an aperture with the same radius as the interaction cylinder, multiplied by a term that depends on the phase mismatch due to the angle φ.

The output angle in far-infrared difference-frequency generation can be larger than the input angle, and this property has been used to separate the input and the output frequencies spatially.[173]

2.17 THE ELECTROOPTIC COEFFICIENT

By treating the electrooptic effect as a mixing of two waves with frequencies ω_1 and zero to create a third wave at ω_1, as we have done before, we can use the coupled amplitude equations to find the relation between the normally used electrooptic coefficient r_{ij} and the nonlinear susceptibility d_{ij}. Taking $\mathscr{E}_1 e^{-i\omega t}$ as the light wave, $\mathscr{E}_2$ as the applied field, and $\mathscr{E}_3 e^{-i(\omega t - \varphi)}$ as the generated wave, we find from the last expression of equation 2.39

$$\frac{d\mathscr{E}_3}{dz} = -i \frac{2\pi\omega^2}{kc^2} d \, \mathscr{E}_1 \mathscr{E}_2$$

and, writing out the frequency dependence, this gives for a crystal of length L

$$\mathscr{E}_3 e^{-i(\omega t - \varphi)} = -i \frac{2\pi\omega^2}{kc^2} Ld \, \mathscr{E}_1 \mathscr{E}_2 e^{-i\omega t}$$

and since $-i = e^{-i\pi/2}$, this can be written as

$$\mathscr{E}_3 e^{-i(\omega t - \varphi)} = \frac{4\pi^2 L}{n\lambda} d \, \mathscr{E}_1 \mathscr{E}_2 e^{-i[\omega t + (\pi/2)]}$$

In other words, we have generated a small-amplitude $\mathscr{E}_3$ which leads 90° in phase. Now, using the normal electrooptic coefficient, we find that an applied field $\mathscr{E}_2$ produces a change in refractive index. We have

$$\Delta\left(\frac{1}{n^2}\right) = r\mathscr{E}_2$$

and so

$$\Delta n = \frac{-n^3 r \mathscr{E}_2}{2}$$

In a crystal of length L this index difference will produce a phase change $\Delta\varphi = -\pi n^3 r \mathscr{E}_2 L/2$, which should be the same as the one produced by the effect on $\mathscr{E}_1$ of the small $90°$ out-of-phase component $\mathscr{E}_3$. Since $\mathscr{E}_3 \ll \mathscr{E}_1$, we can write for this last phase change

$$\Delta\varphi = \tan^{-1}\frac{|\mathscr{E}_3|}{|\mathscr{E}_1|} \approx \frac{|\mathscr{E}_3|}{|\mathscr{E}_1|}$$

and so we have

$$-\frac{\pi n^3 r \mathscr{E}_2 L}{\lambda} = \frac{4\pi^2 L d \mathscr{E}_2}{n\lambda}$$

or

$$r = -\frac{4\pi}{n^4} d$$

From equation 2.20 we can find, by contraction of the appropriate indices, that the electrooptic matrix can be determined from the nonlinear susceptibility matrix by interchanging rows and columns. Thus we can write

$$r_{ij} = -\frac{4\pi}{n^4} d_{ji} \tag{2.77}$$

2.18 NONLINEAR INTERACTIONS IN REFLECTION

In Chapter 1 we learned that the rectilinear propagation of light in a medium is due to coherent scattering in the forward direction. In any other direction the different radiating dipoles are out of phase and interfere destructively. At the boundary of the material, however, the radiating dipoles on one side of the surface are different from those on the other side; as a result, some of the radiation is coherently scattered in a backward direction. This is the well-known Fresnel reflection at the boundary of a medium. Similarly, if the polarization of the material contains higher harmonics and mixed frequencies, these frequencies may also be present in the reflected radiation. However, the angle of incidence will be equal to the angle of reflection for the linear reflected wave only—the nonlinear waves may be reflected at different angles. This effect was predicted by Bloembergen and Pershan[22] and was experimentally verified by Ducuing and Bloembergen.[49]

2.19 DIMENSIONS

Except where specifically indicated, the c.g.s. system is used throughout this book. The dimension of the nonlinear coefficient in this system is centimeters per stat. volt. In MKS units this becomes meters per volt. The conversion from c.g.s. to MKS units is given by

$$d \text{ (MKS)} = \frac{4\pi}{3 \times 10^4} d \text{ (esu)}$$

In the MKS system the nonlinear polarization is given by

$$P = \varepsilon_0 \, d \, EE$$

where ε_0 is the permittivity of free space. Some authors include ε_0 in the coefficient. Then the conversion from electrostatic units to MKS units becomes $d \text{ (MKS)} = 3.68 \times 10^{-15} d \text{ (esu)}$.

3

Phase Matching

3.1 INTRODUCTION

The term $(\sin x/x)^2$ in equations 2.44 to 2.46 is crucial to the success of any frequency-mixing experiment. As mentioned before, it is a measure of the phase mismatch between the polarization wave and the generated electromagnetic wave, and because of normal color dispersion this mismatch is usually large. To obtain a significant amount of power, it is obviously necessary to achieve, somehow, a condition where the phases of these two waves are matched. In this chapter we first look at the result of the mismatch in more detail, and then we introduce several methods used to obtain phase matching.

3.2 POWER FLOW IN THE NON-PHASE-MATCHED CASE

To illustrate the effects of the phase mismatch let us examine the diagrams of Figure 3.1. Here we have a number of discrete oscillators spaced at exactly one full wavelength of the polarization wave. The wavelength of the electromagnetic wave is assumed to be equal to 8/9 of this spacing. In other words $\Delta k = \pi/4$ per spacing. The figures show the phases and amplitudes of three waves: O, the wave generated by the oscillator; P, the resultant of the waves generated by the previous oscillators; and R the resultant of O and P. In Figure 3.1a the only wave present is O, but in Figure 3.1b the same wave, now labeled P, is $\pi/4$ behind in phase compared with the wave O generated by the second oscillator. The resultant of these two waves is R. In Figure 3.1c the same wave R is labeled P and is $3\pi/8$ behind in phase compared with the wave generated by the third oscillator. This results in R in Figure 3.1c, and so on. However, when $\Delta k \cdot L = \pi$ (i.e., after four spacings), the amplitude of P has reached its maximum. After four more spacings the wave generated by the ninth oscillator exactly cancels the wave generated by the first eight oscillators, resulting in no signal at all. The process then starts all over again, clearly it is periodic; as the thickness of the crystal increases, the output

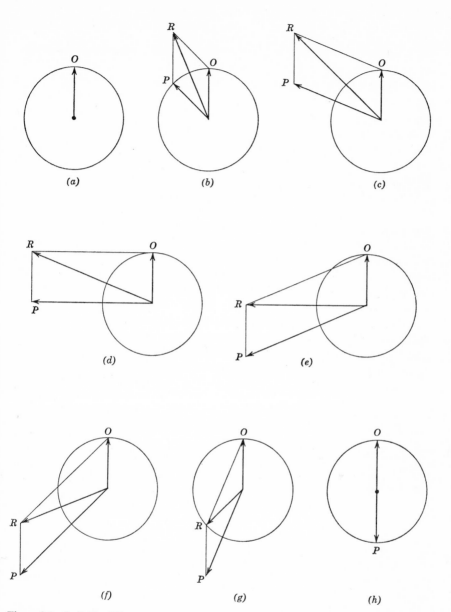

Figure 3.1 Individual illustrations of the effects of a phase mismatch between the polarization wave and the generated electromagnetic wave.

signal alternately increases and goes to zero. This of course follows also from the previous equations (2.44–2.46), since $L^2 (\sin \Delta k L/2)^2/(\Delta k L/2)^2$ varies as $\sin^2$ when the value of L changes. The crystal length at which the signal reaches its first maximum (Figure 3.1e) is usually known as the coherence length, L_{coh}. Clearly, if the crystal is not phase matched, the most we can ever expect to generate will be the signal from one coherence length, no matter how long the crystal is. If, on the other hand, $\Delta k = 0$, the signal will be proportional to the square of the length of the crystal, at least in the small-signal approximation. In other words, *if the crystal is phase matched the power generated is large; if not, only very little power is obtained.*

The question now is, what happens to the power when the crystal is not phase matched? Or, more specifically, what happens to the power used to generate those frequency components of the nonlinear polarization for which the crystal is not phase matched? The question is important, since these frequency components are always present. Thus if this power were lost (e.g., by absorption), any nonlinear generation would always be a lossy interaction.

The answer to this question is found in the phase of the signal. Neglecting the tensor aspect of ε, we find from equation 1.17 the energy density flux per unit volume. Using script letters specifically to indicate real quantities we write

$$\nabla \cdot \mathscr{S} = \frac{1}{4\pi}\left[\mathscr{E}\frac{\partial \mathscr{E}}{\partial t} + \mathscr{H}\frac{\partial \mathscr{H}}{\partial t} \right] + \mathscr{E}\frac{\partial \mathscr{P}}{\partial t} \tag{3.1}$$

the bracketed term on the right-hand side of equation 3.1 is the rate of increase of the vacuum electromagnetic energy, and the last term is the power spent in varying the electric polarization. Per unit volume, the power input to the polarization is

$$W_{\mathrm{vol}} = \overline{\mathscr{E} \cdot \frac{\partial \mathscr{P}}{\partial t}}$$

where the horizontal bar indicates averaging over a period that is long compared with $2\pi/\omega$. Thus the average power absorption is

$$W_{\mathrm{vol}} = \tfrac{1}{2}\omega \mathscr{P}\mathscr{E} \sin \varphi \tag{3.2}$$

where φ is the phase difference between the polarization wave and the electromagnetic wave.

As we saw in Section 2.17, for $\Delta k = 0$ the generated signal is 90° out of phase with the polarization wave. This means that W_{vol} is negative (i.e., the power is coupled from the polarization wave into the electromagnetic wave). For $\Delta k \neq 0$, however, this 90° phase lag exists only at $L = 0$, and after one coherence length the phase of the signal has changed by exactly 90°. As a result, the power flow changes sign; instead of power being coupled from the polarization wave into the electromagnetic wave, it is now coupled from the electromagnetic wave into the polarization wave. In other words, the power is coupled back into the input waves.

Thus we find that, if the interaction is not phase matched, the power shuttles back and forth between the generated wave and the input waves. In second-harmonic generation, for example, the input at ω_1 will generate a wave at $2\omega_1$ in the first coherence length, but in the second coherence length the wave at $2\omega_1$ will produce a difference frequency with the wave at ω_1, to give a wave at $2\omega_1 - \omega_1 = \omega_1$. This means that if the crystal is exactly two coherence lengths long, no second harmonic will be generated, but also no power will be lost from the fundamental. Thus in general, if only one frequency is phase matched, all the power lost from the fundamental is coupled into this frequency, and no power is lost to any of the non-phase-matched components.

The variation of the second-harmonic signal with crystal length was shown first by Terhune and his co-workers, who varied the effective thickness of a quartz crystal by rotating it. Their results are presented in Figure 3.2. The

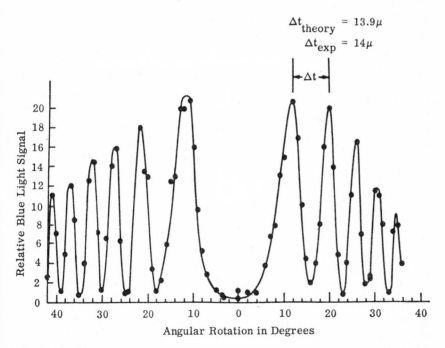

Figure 3.2 Variation of the output signal when the crystal is rotated. A similar figure results as a function of temperature. After Maker et al., *Phys. Rev. Lett.*, **8**, 21 (1962).

distance between the successive maxima corresponds to 14 μ, and the value of twice the coherence length calculated from refractive index data is 13.9 μ.

3.3 QUASI-PHASE-MATCHING METHODS

If we could change the difference in phase between the polarization wave and the electromagnetic wave by $\pi/2$ every time the length of the crystal increases by one coherence length, we would obtain a quasi-phase-matched condition. It would be quasi-phase matched because the signal from one coherence length of phase-matched crystal is $\pi^2/4$ larger than the signal from the same length of non-phase-matched crystal, since $(\sin x/x)^2 = 4/\pi^2$ for $x = \pi/2$. Therefore, the signal from our hypothetical crystal would still not

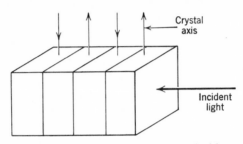

Figure 3.3   Quasi phase matching in a stack of plates rotated with respect to one another.

be quite as large as the signal from a phase-matched crystal with the same d_{eff} and length. A way to do this was suggested by Bloembergen et al.[5] It consists of making thin plates of the crystal, one coherence length thick, and turning alternate platelets over, so that the polarization wave will undergo a 180° phase change going from one plate to the next (Figure 3.3). The experimental problems of the approach are obvious; for quartz, for example, all the plates would have to be exactly 7 μ thick and all would have to be in optical contact.

In a more recent paper, Bloembergen and Sievers[24] proposed to overcome this drawback by growing semiconductor layers epitaxially onto one another. This kind of artificial crystal shows interesting properties which do not occur in the bulk form of the separate constituents. A discussion of these properties falls outside the scope of this book.

A different method to correct the phase mismatch periodically was also suggested by Bloembergen and his co-workers,[5] and was experimentally verified by Ashkin et al.[8] and by Boyd and Patel.[27] It uses the phase change on total internal reflection. Both the fundamental and the second harmonic are reflected between the top and the bottom surface of a slab of crystal. The angle of reflection is so chosen that the phase mismatch accumulated in every pass between the two reflecting sides is just canceled by the differential phase change between the fundamental and the second-harmonic reflections. Recently this approach was used to generate the second harmonic of a carbon dioxide laser using a thin piece of gallium arsenide as an optical waveguide.[3]

The advantage of all these quasi-phasematching methods over the methods described in Section 3.4 is that they can be used in isotropic materials.

3.4 ANGLE PHASE MATCHING

A method to obtain true phase matching (i.e., to make $\Delta k = 0$) was published by Terhune and co-workers[105] and, independently, by Giordmaine[65] in 1962. The approach employs the birefringence of a uniaxial crystal.

To explain how this method works, let us take KDP as an example. The refractive indices of KDP are shown in Figure 3.4. The crystal is negative

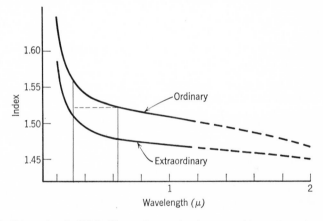

Figure 3.4 Dispersion in KDP, illustrating angle phase matching. The ordinary index at 6328 Å is larger than the extraordinary index at 3164 Å, making phase matching possible for second harmonic generation with 6328 Å as the fundamental.

uniaxial, so the ordinary index is larger than the extraordinary index. To obtain collinear phase matching for second-harmonic generation, we want to have equal refractive indices for the second-harmonic and the fundamental frequencies. Let us assume that a helium–neon laser at 6328 Å is used as a source. From Figure 3.4 we observe that the ordinary index at 6328 Å is larger than the extraordinary index at 3164 Å. However, as we have seen in Section 1.4, the index for an extraordinary ray can be varied by changing the angle between the wave normal and the optic axis. Thus we should be able to transmit the wave at an angle θ to the optic axis, such that the refractive

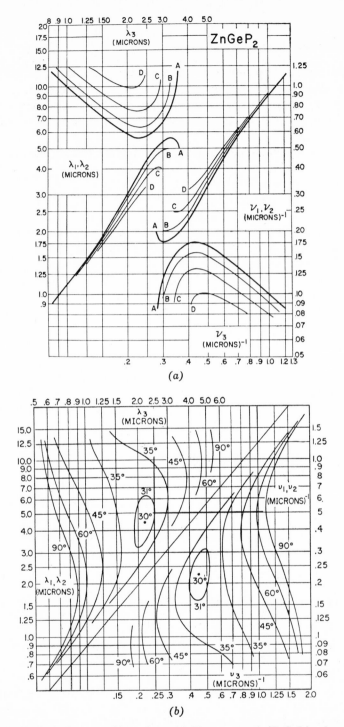

Figure 3.5 Typical phase-matching plots: (*a*) For a crystal (ZnGeP$_2$) where phase-matched second harmonic generation is not possible. After Boyd et al., *Appl. Phys. Lett.*, **18**, 301 (1971); (*b*) for a crystal (AgGaS$_2$) where phase matching is possible for second harmonic generation. After Boyd et al., *IEEE J. Quant. Electr.*, **QE-7**, 563 (1971).

index for the input, polarized as an o-ray, is exactly equal to the refractive index for the second harmonic polarized as an e-ray.

Obviously this can be done for all those wavelengths λ for which the ordinary index is larger than the extraordinary index at $\lambda/2$. The angle θ can be calculated from equation 1.26, recalling that, for a precise calculation of this angle, the indices to be used are the indices of the material with respect to a vacuum.

In this example we have used ordinary rays as fundamentals and an extraordinary ray as the second harmonic. For positive uniaxial crystals this procedure would have been reversed—extraordinary fundamentals and ordinary second harmonic—and we might even mix the polarization of the two input beams and have an ordinary ray and an extraordinary ray as fundamentals, producing an ordinary second harmonic in a positive crystal and an extraordinary second harmonic in a negative crystal. If both fundamentals have the same polarization, we speak of type I phase matching, and if their polarizations are orthogonal the phase matching is called type II. It should also be clear that, although we used second-harmonic generation in the example, this phase-matching method would work equally well if the two fundamentals did not have the same frequency.

Figure 3.4 is instructive in understanding angle-tuned phase matching, but it does not given information about the values of the phase-matching angles necessary for different interactions. This can be done by plotting the output frequency ν_3 on the x axis and the input frequencies ν_1 and ν_2 on the y axis of a rectangular coordinate system. Each output frequency can of course be generated by many different combinations of inputs, for we can write $\nu_3 = [\frac{1}{2}\nu_3 - \Delta\nu] + [\frac{1}{2}\nu_3 + \Delta\nu]$. For a given phase-matching angle, only one combination is possible, yielding two values of y for every value of x. If all the points for a given phase-matching angle are connected, two curves result in the general case; if phase-matched second-harmonic generation is possible, these two curves degenerate into one. Figure 3.5 presents typical examples for both cases. Note that frequencies are shown in one half and wavelengths in the other half of each diagram.

3.5 THE EXPRESSION FOR d_{eff} FOR THE DIFFERENT CRYSTAL CLASSES

We have blithely assumed that it is possible to couple to the desired polarization of the second harmonic for the given polarizations of the fundamentals. Whether this can indeed be done depends on the particular symmetry of the crystal, and as we shall see, for some crystal classes it is impossible. We now examine the different crystal classes that lack a center of symmetry to see which ones allow phase matching in this manner.

Quite generally, we have

$$\omega_1 + \omega_2 = \omega_3 \tag{3.3}$$

and by setting $\Delta k = 0$ in equation 2.40 we find

$$k_1 + k_2 = k_3 \tag{3.4}$$

If we solve equation 2.37, keeping track of all the spatial components, we find in the more general case that

$$\mathbf{k}_1 + \mathbf{k}_2 = \mathbf{k}_3 \tag{3.5}$$

It is easily seen that these equations mean that for efficient conversion both energy (equation 3.3) and momentum (equation 3.5) have to be conserved.

Since we are primarily interested in the case where all three beams are collinear, we use equation 3.4. From this we find

$$n_1\omega_1 + n_2\omega_2 = n_3\omega_3$$

We can find the components along the x, y, and z axes of an ordinary wave propagating at an angle θ to the z axis in a plane at an angle φ to the x axis by multiplying the amplitude of the wave by the appropriate direction cosines

$$E_j{}^o = |\mathbf{E}^o| \begin{vmatrix} \sin \varphi \\ -\cos \varphi \\ o \end{vmatrix} = |\mathbf{E}^o| \cdot a_j$$

For an extraordinary wave this becomes

$$E_j{}^e = |\mathbf{E}^e| \begin{vmatrix} -\cos \varphi \cos \theta \\ -\sin \varphi \cos \theta \\ \sin \theta \end{vmatrix} = |\mathbf{E}^e| \cdot b_j$$

If we use these components to construct the vector $\mathbf{F}$ of equation 2.22 and then use this vector in equation 2.23, we find the components P_i of a polarization vector $\mathbf{P}$. From these components, in turn, we obtain the effective polarization seen by an ordinary- or extraordinary wave in the direction (φ, θ) by performing the following summations:

$$P_{\text{eff}}^o = a_i P_i$$

$$P_{\text{eff}}^e = b_i P_i$$

Thus for type I phase matching in a negative uniaxial crystal, using the noncontracted notation, we find

$$|\mathbf{P}^e| = (2 - \delta_{jk}) b_i d_{ijk} a_j a_k |\mathbf{E}^o|^2 \tag{3.6}$$

For the other cases we find

$$|\mathbf{P}^o| = (2 - \delta_{jk})a_i d_{ijk} b_j b_k |\mathbf{E}^e|^2 \qquad \text{type I, positive} \qquad (3.7)$$

$$|\mathbf{P}^e| = (2 - \delta_{jk})b_i d_{ijk} a_j b_k |\mathbf{E}^o| |\mathbf{E}^e| \qquad \text{type II, negative} \qquad (3.8)$$

$$|\mathbf{P}^o| = (2 - \delta_{jk})a_i d_{ijk} a_j b_k |\mathbf{E}^o| |\mathbf{E}^e| \qquad \text{type II, positive} \qquad (3.9)$$

We can now derive the expressions for d_{eff} for the four cases shown in equations 3.6 through 3.9 for any of the 13 uniaxial crystal classes lacking a center of symmetry. The results of this derivation appear in Table 3.1. For each class we find four expressions: two for an interaction between two e-rays and an o-ray, and two for an interaction between two o-rays and an e-ray.

Table 3.1 gives the most general set of equations for birefringent phase matching. Four different cases are distinguished; two o-rays producing an e-ray, two e-rays producing an o-ray, an e-ray and an o-ray producing an e-ray, and an e-ray and an o-ray producing an o-ray. These four cases show a definite difference only in those interactions where the frequency region of interest contains an absorption band.

On the other hand, whenever all three frequencies lie in a region without absorption, Kleinman's symmetry condition is valid and it makes no difference which of the three waves is the output. That is, if Kleinman's condition is taken into account, we find one expression for an interaction between two o-rays and an e-ray and one for an interaction between two e-rays and an o-ray. The expressions for d_{eff} in the cases in which Kleinman's symmetry condition is valid are listed in Table 3.2.

It is worthwhile to study Tables 3.1 and 3.2 thoroughly, since a number of interesting points become evident. First of all, these tables, and also Figure 2.1, indicate that the expressions for d_{eff} for some crystal classes can be derived from those for other crystal classes. For example, class $\bar{6}$ combined with class 6 gives class 3, and class $\bar{6}$m2 plus class 6mm gives class 3m.

Another interesting point is that for some crystal classes the angle with respect to the x and y axes is important, but in other classes this angle does not enter into the expression for d_{eff} at all. It is even more significant to note that in some crystal classes only one part of the expression is dependent on the angle φ. In class 3m, for example, for the case of mixing between two o-rays and one e-ray we find that the term containing d_{22} vanishes when $\sin 3\varphi$ is equal to 0. In other words, if the phase matching is done in the xz plane ($\varphi = 0$), then only the d_{15} term is measured. This property can be used to measure the sign of the two coefficients d_{15} and d_{22}. In lithium niobate, for example, second-harmonic generation in a crystal cut at an angle for which $\sin 3\varphi$ has a negative value gave a smaller input than the same interaction in a similar crystal cut at an angle for which $\sin 3\varphi$ is zero. From this result it was concluded that the coefficients d_{11} and d_{22} in lithium niobate

TABLE 3.1 EQUATIONS TO BE USED FOR d_{eff} FOR THE 13 UNIAXIAL CRYSTAL CLASSES FOR CASES WHERE KLEINMAN'S SYMMETRY DOES NOT HOLD: (a) TWO e RAYS AND ONE o RAY

Crystal Class	Type I	Type II
6 and $\bar{4}$	$-d_{14}\sin2\theta$	$d_{14}\sin\theta\cos\theta$
622 and 422	$-d_{14}\sin2\theta$	$d_{14}\sin\theta\cos\theta$
6mm and 4mm	0	0
$\bar{6}$m2	$d_{22}\cos^2\theta\cos3\phi$	$d_{22}\cos^2\theta\cos3\phi$
3m	$d_{22}\cos^2\theta\cos3\phi$	$d_{22}\cos^2\theta\cos3\phi$
$\bar{6}$	$\cos^2\theta(d_{11}\sin3\phi+d_{22}\cos3\phi)$	$\cos^2\theta(d_{11}\sin3\phi+d_{22}\cos3\phi)$
3	$\cos^2\theta(d_{11}\sin3\phi+d_{22}\cos3\phi)$ $-d_{14}\sin2\theta$	$\cos^2\theta(d_{11}\sin3\phi+d_{22}\cos3\phi)$ $+d_{14}\sin\theta\cos\theta$
32	$d_{11}\cos^2\theta\sin3\phi-d_{14}\sin2\theta$	$d_{11}\cos^2\theta\sin3\phi+d_{14}\sin\theta\cos\theta$
$\bar{4}$	$d_{14}\sin2\theta\cos2\phi$ $-d_{15}\sin2\theta\,\sin2\phi$	$(d_{14}+d_{36})\sin\theta\cos\theta\cos2\phi$ $-(d_{15}+d_{31})\sin\theta\cos\theta\sin2\phi$
$\bar{4}$2m	$d_{14}\sin2\theta\cos2\phi$	$(d_{14}+d_{36})\sin\theta\cos\theta\cos2\phi$

TABLE 3.1(b) TWO *o* RAYS AND ONE *e* RAY

Crystal Class	Type I	Type II
6 and 4	$d_{31}\sin\theta$	$d_{15}\sin\theta$
622 and 422	0	0
6mm and 4mm	$d_{31}\sin\theta$	$d_{15}\sin\theta$
$\bar{6}$m2	$-d_{22}\cos\theta\sin3\phi$	$-d_{22}\cos\theta\sin3\phi$
3m	$d_{31}\sin\theta-d_{22}\cos\theta\sin3\phi$	$d_{15}\sin\theta-d_{22}\cos\theta\sin3\phi$
$\bar{6}$	$\cos\theta(d_{11}\cos3\phi-d_{22}\sin3\phi)$	$\cos\theta(d_{11}\cos3\phi-d_{22}\sin3\phi)$
3	$\cos\theta(d_{11}\cos3\phi-d_{22}\sin3\phi)+d_{31}\sin\theta$	$\cos\theta(d_{11}\cos3\phi-d_{22}\sin3\phi)+d_{15}\sin\theta$
32	$d_{11}\cos\theta\cos3\phi$	$d_{11}\cos\theta\cos3\phi$
$\bar{4}$	$-\sin\theta(d_{31}\cos2\phi-d_{36}\sin2\phi)$	$-\sin\theta(d_{15}\cos2\phi+d_{14}\sin2\phi)$
$\bar{4}$2m	$-d_{36}\sin\theta\sin2\phi$	$-d_{14}\sin\theta\sin2\phi$

TABLE 3.2 EQUATIONS FOR d_{eff} FOR CASES WHERE KLEINMAN'S SYMMETRY IS VALID

Crystal Class	Two e rays and one o ray	Two o rays and one e ray
6 and 4	0	$d_{15}\sin\theta$
622 and 422	0	0
6mm and 4mm	0	$d_{15}\sin\theta$
$\bar{6}$m2	$d_{22}\cos^2\theta\cos\phi$	$-d_{22}\cos\theta\sin3\phi$
3m	$d_{22}\cos^2\theta\cos3\phi$	$d_{15}\sin\theta-d_{22}\cos\theta\sin3\phi$
$\bar{6}$	$\cos^2\theta(d_{11}\sin3\phi+d_{22}\cos3\phi)$	$\cos\theta(d_{11}\cos3\phi-d_{22}\sin3\phi)$
3	$\cos^2\theta(d_{11}\sin3\phi+d_{22}\cos3\phi)$	$d_{15}\sin\theta+\cos\theta(d_{11}\cos3\phi-d_{22}\sin3\phi)$
32	$d_{11}\cos^2\theta\sin3\phi$	$d_{11}\cos\theta\cos3\phi$
$\bar{4}$	$\sin2\theta(d_{14}\cos2\phi-d_{15}\sin2\phi)$	$-\sin\theta(d_{14}\sin2\phi+d_{15}\cos2\phi)$
$\bar{4}$2m	$d_{14}(\sin2\theta\cos2\phi)$	$-d_{14}\sin\theta\sin2\phi$

have opposite signs.[18] For difference-frequency generation of far infrared in quartz, class 32, it was shown in a similar manner that the coefficient d_{14} is unequal to zero.[173] This is as expected, since Kleinman's symmetry does not apply for this type of difference-frequency generation.

Note that the expressions for d_{eff} given in Tables 3.1 and 3.2 include only those for birefringent phase matching in uniaxial crystals. The expressions necessary for phase matching in biaxial crystals and those for interactions in isotropic crystals can be developed using the same method.

3.6 DISADVANTAGES OF ANGLE PHASE MATCHING

As we have seen in Chapter 1, the ray direction and the wave-normal direction for an extraordinary wave are parallel only when $\theta = 0$ or when $\theta = 90°$. Thus in a phase-matched interaction at an intermediate angle θ, the extraordinary beam does not overlap the ordinary beams in the entire interaction length. For a type I interaction this effect, although present, is not too serious. It only means that the generated beam does not totally overlap the polarization wave, and thus the integration in equation 2.44 becomes more complicated. The exact form of integration has to be worked out for each specific case. In general, we find that the output is proportional not to the square, but rather to a lower power of the length. For a type II interaction the effect is more serious, because here the two fundamental beams do not overlap completely, and thus after a certain crystal length the polarization wave vanishes completely and mixing no longer occurs.

Another limitation of angular phase matching is due to the divergence of a focused beam. We illustrate this for the general case of second harmonic generation. For second-harmonic generation in a negative crystal and a type I interaction, we find from equation 2.40 that

$$\Delta k = \frac{2\omega_1}{c} [n_2^e(\theta) - n_1^o] \tag{3.10}$$

if $\theta = \theta_m$, $\Delta k = 0$; but for a small deviation $\Delta\theta$ from this angle, we find, using equation 1.26 that

$$\Delta k = -\frac{\omega_1}{c} (n_1^o)^3 [(n_2^e)^{-2} - (n_2^o)^{-2}] \sin 2\theta_m \cdot \Delta\theta \tag{3.11}$$

Similarly, for a positive crystal and a type I process, we find that

$$\Delta k = -\frac{\omega_1}{c} (n_2^o)^3 [(n_2^e)^{-2} - (n_2^e)^{-2}] \sin 2\theta_m \cdot \Delta\theta \tag{3.12}$$

For type II interactions it is easily shown that the mismatch is reduced by a factor of 2. For crystals with a small birefringence and dispersion, these expressions reduce to

$$\Delta k \approx \beta \cdot \frac{2\omega}{c} (n^{\text{o}} - n^{\text{e}}) \sin 2\theta_{\text{m}} \cdot \Delta\theta$$

where β has the same sign as the crystal birefringence and is equal to 1 for type I and $\frac{1}{2}$ for type II interactions. Thus we see that the variation of Δk is linear with $\Delta\theta$. In practice this can cause difficulties for intermediate angles θ_{m}. The output is proportional to the energy density at the fundamental frequency, and so, to achieve the maximum energy density, the beam is focused on the crystal. However, the linear variation of Δk with $\Delta\theta$ means that, for a given convergence of the beam, efficient phase matching will be obtained over a restricted crystal length only. (See also Chapter 5.)

3.7 TEMPERATURE-DEPENDENT PHASE MATCHING

The disadvantages mentioned in the last section can be minimized by making the matching angle θ_{m} equal to $90°$. The walk-off of the extraordinary ray is then nonexistent. From equations 3.9 and 3.10 we see that, since for values of θ_{m} close to $90°$ we have $(\sin 2\theta_{\text{m}}) \Delta\theta = 2(\Delta\theta)^2$, we can write

$$\Delta k = \frac{2\omega_1}{c} (n_1^{\text{e}})^3 [(n_2^{\text{o}})^{-2} - (n_2^{\text{e}})^{-2}](\Delta\theta)^2$$

for positive crystals. The expression for negative crystals is

$$\Delta k = \frac{2\omega_1}{c} (n_2^{\text{e}})^3 [(n_2^{\text{e}})^{-2} - (n_2^{\text{e}})^{-2}](\Delta\theta)^2$$

For type II interactions, the mismatches are again reduced by a factor of 2. Thus we see that for $\theta_{\text{m}} = 90°$ the allowable divergence of the beams is much larger. In addition, there are no first-order walk-off effects due to double refraction.

For these reasons, ninety degree phase matching is often called noncritical phase matching. It can be obtained in some crystals by varying the temperature of the crystal; the extraordinary index is in general much more temperature dependent than the ordinary index. Thus by changing the temperature it is possible to change the birefringence until phase matching is obtained for $\theta = 90°$.

In difference-frequency generation for cases of the difference frequency lying on the other side of a *reststrahlen* band, temperature-dependent phase matching may even be possible for the case of a nonbirefringent crystal. A variation of the temperature shifts the position of the absorption edge and therefore changes the values of the refractive indices for the two input frequencies, whereas the refractive index for the difference frequency changes relatively little. This was demonstrated in InSb using 1040 and 940 cm^{-1} as the respective input frequencies, to obtain an output at 100 cm^{-1}. Phase matching was observed at a temperature of $-40°C$.[174]

The exact temperature for 90° phase matching can be varied by changing the chemical composition of the crystal. This effect was discovered more or less in reverse when it was noted that different crystals of lithium niobate showed different phase-matching temperatures for the same interaction. (See Chapter 4.) For difference frequency generation the effect of tuning by changing the chemical composition of the crystal was described by Nguyen Van Tran et al.[158]

3.8 PHASE MATCHING IN BIAXIAL CRYSTALS

The phase matching described in Section 3.4 for uniaxial crystals is dependent only on the angle θ between the wave normals and the z axis. If the angle φ of the plane of incidence with respect to the x axis is changed, the value of θ remains constant. This is because the wave surface of the ordinary ray is a sphere and the wave surface of the extraordinary ray is an ellipsoid of revolution. The intersection between these two wave surfaces is a circle which is centered on the optic axis and which can be thought of as the base of a cone with vertex 2θ. The optic axis, moreover, coincides with the z axis. This is the case of type I phase matching. For type II phase matching we similarly find a circular cone whose base is the intersection between the extraordinary ellipsoid of revolution and an ellipsoid of revolution that represents $\frac{1}{2}(n_o + n_e)$. Again, the angle θ is independent of the angle φ.

For biaxial crystals the wave surfaces do not simply form a sphere and an ellipse, they form a more complicated, two-sheeted surface. There are now two optic axes, and neither one coincides with the z axis. However, the phase-matching directions still lie in a cone around the optic axes, but this cone does not necessarily have a circular cross section. This can be understood as follows. Imagine two cones with circular cross section, each centered on one optic axis, but having a vertex that is larger than the angle between the two optic axes. Here the cones include the z axis and form one cone, as it were, which does not have a circular cross section. The exact form of these cones depends on the sequence of the different refractive indices. The different possible cases have been examined in detail by Hobden.[78]

We give as an example only the case where $n_x(\omega_2) > n_x(\omega_1) > n_y(\omega_2) > n_y(\omega_1) > n_z(\omega_2) > n_z(\omega_1)$, and we examine this case only for type I phase matching. To find the phase-matching directions we draw the two surfaces shown in Figure 3.6. Note how the polarizations of the fundamental and the second-harmonic interchange as we go around the intersection of the two

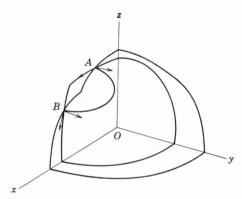

Figure 3.6 Illustration of phase matching in biaxial crystals.

surfaces. In A the fundamental is polarized parallel to the x–z plane and the second harmonic is polarized perpendicular to this plane; but in B the fundamental is polarized perpendicular to the x-z plane, and the second harmonic is polarized in the x-z plane. By referring to the polarizations in Figure 1.8, we see that the two polarizations rotate as we move around the intersection. Clearly in biaxial crystals the angle θ is not independent of the angle φ.

To find a mathematical solution for the phase matching directions, we return to equation 1.24 and write this equation for both frequencies. The locus of the phase-matching directions is then found by solving the two equations simultaneously. It is not possible to obtain a solution in closed form, and these phase-matching problems are best solved by computer calculation.

3.9   OTHER PHASE-MATCHING METHODS

The type of phase matching featuring the birefringence of the crystal is the one most universally employed. However, a variety of other methods have been used, such as matching in optically active media[15,138] and matching using Faraday rotation.[134] In more recent work optical waveguides have been used to reduce the mismatch[156] and phase matching has been induced acoustically.[127]

In those cases where $k_1 + k_2 > k_3$, phase matching can be achieved by making the interaction take place between noncollinear beams.[157,175] If the difference in frequency between the inputs and the outputs is large, large angles can occur between the beams inside the crystal; moreover if the crystal has a high refractive index, the angles can become large enough to allow total internal reflection at the exit face to occur. A typical case is illustrated in Figure 3.7, where the two inputs were 1080 and 945 cm^{-1}. The figure reveals

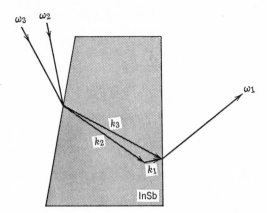

Figure 3.7 Phase-matched interaction for non-collinear beams.

that the crystal was ground in the shape of a prism to allow the output to couple out.

3.10 COMPETING INTERACTIONS

As mentioned before, normally only one of the frequency components of the nonlinear polarization is phase matched, and no power is lost to other frequencies. In some cases, however, it is possible to have simultaneous phase matching for several interactions. A number of frequencies are then generated efficiently, and thus not all the power is available for the one desired frequency. In an extreme case, Yarborough and Amman[169] have shown that optical parametric oscillation, second-harmonic generation, and difference-frequency generation can occur simultaneously in lithium niobate when 1.06 μ is used as the pump and when the crystal has the correct temperature and orientation. In their particular case the optical parametric oscillator gave an idler at 0.405 μ^{-1} and a signal at 0.535 μ^{-1} from a pump at 0.94 μ^{-1} [1.065 μ]. At the same time, the signal was phase matched to give a second harmonic at 1.07 μ^{-1}, and this second harmonic interacted with the idler to give a difference frequency at 0.665 μ^{-1}. The last interaction is also phase matched. If this process were used to create, for example, 0.535 μ^{-1}, the second-harmonic generation and the difference-frequency generation would of course be considered as losses. On the other hand, it is to be noted that the second-harmonic generation converts power to a wavelength shorter than the pump. (See also Chapter 7.)

From Figure 3.5b we can see that simultaneous phase matching for a parametric oscillator and for second-harmonic generation can occur in other parametric oscillators, too. For example, the curve for phase matching at

$35°$ shows that the second harmonic of about $0.31 \ \mu^{-1}$ is phase matched at the same time as the sum frequency generation between about 0.31 and 0.1 μ^{-1}. In other words, if a second harmonic with a pump at $0.41 \ \mu^{-1}$ were used to give an idler at $0.1 \ \mu^{-1}$ and a signal at $0.31 \ \mu^{-1}$, the signal would at the same time be phase matched to give a second harmonic at $0.62 \ \mu^{-1}$.

Another limiting side effect is the production by the generated second harmonic of photoinduced carriers that absorb the pump radiation. This effect has been observed in tellurium.[58]

4

Nonlinear Materials

4.1 HISTORICAL INTRODUCTION

The history of nonlinear materials is either brief or long, depending on one's terms of reference. The first studies of the linear electrooptic effect were made before the start of the twentieth century. However, if we confine our interest to materials for second-harmonic generation (SHG) and other optical frequency-mixing processes, then the first marker in time appears as recently as 1961 when SHG was first observed.[55] To understand the evolution of the different requirements for nonlinear optical materials, it is instructive to follow the historical development.

From basic theoretical considerations it was known, before any experiments were performed, that the crystal should be noncentrosymmetric and that the symmetry requirements for SHG and for piezoelectricity were the same. Also, an obvious requirement was that the crystal be transparent for all frequencies in the interaction. A material that is readily available in large, perfect, single crystals and, moreover, satisfies the requirements just named in the visible and the near-infrared is quartz. Thus Franken and his co-workers used quartz when they made the first experimental observation of SHG.[55] Their source was a Q-switched ruby laser, and the second harmonic was detected by the use of filters and a spectrograph. The signal was very small, but it started an era. In retrospect, we can see why the signal was so small: not only is the nonlinearity of quartz very small, but also, and more important, the generation was not phase matched.

The second major advance came with the discovery of phase matching (see Chapter 3) published simultaneously in two papers by Maker et al.[105] and Giordmaine.[65] Both used the crystal ammonium-dihydrogen phosphate (ADP), and both reported huge gains in SHG efficiency over the first work.

Using this new phase-matching technique, SHG efficiencies approaching 10 to 50% were soon reported using pulsed lasers, and even the much less powerful visible and near infrared lines from a helium–neon laser (6328 Å and 1.15 μ) were doubled in frequency. Thus a new requirement for a useful nonlinear optical material was added to the two stated earlier—namely, that the crystal be birefringent, with a large enough birefringence to allow phase matching. A glance at the tables in Appendix 2 shows that some of the materials with the highest nonlinear coefficients (e.g., gallium arsenide) are not birefringent. Thus, their value in nonlinear optical applications is limited, and therefore, as we discuss later in this section, much of the most recent materials development work has centered on efforts to develop birefringent materials with the high nonlinearity typical of gallium arsenide.

The condition that the birefringence be large enough to allow phase matching rules out quartz, which is birefringent but not sufficiently so. Yet quartz, because of its ready availability and its good optical properties, is often used as a reference material against which other materials are measured in a measuring setup in which phase matching is not necessary.

Another fairly obvious requirement that was recognized from the start was that the crystals must have excellent optical quality. This means that for new materials, for which single crystal specimens are not available, it is necessary to grow single crystals of good optical quality. Thus in many cases the search for new and better nonlinear optical material was very largely a crystal-growing effort. It was realized later that the requirements on optical quality for a useful nonlinear material are more stringent than even the most exacting requirements on optical quality for materials used in linear optics.

Taking all the requirements together made an impressive (depressive!) list, supplying few clear rules for how to meet these needs. Perhaps the most obvious approach was to search for promising combinations among all the available data on optical crystalline materials for refractive index, transmission, and crystal class. Such information had been tabulated in convenient form by Winchell and Winchell,[165] Dana and Ford,[46] and Larsen and Berman.[98] Out of this work came several new candidates for nonlinear materials, such as proustite (Ag_3AsS_3),[80] cinnabar (HgS),[30] and potassium dithionate ($K_2S_2O_6$).[76] Another approach was to search all the crystals developed for piezoelectric transducer work during the period 1940 to 1950. This yielded some materials of considerable academic interest but, as yet, none that would be appropriate for devices.[78]

Some greater order was injected into the search when Miller[116] propounded an empirical rule that noted a close relation between linear polarizability (refractive index) and nonlinear polarizability. Prior to this, there had been some attempts to derive theoretical relations to predict the nonlinear coefficients, but expressions that led to useful predictions had not been produced. Miller's empirical rule (Section 2.7) represented the first confirmation of another generally accepted requirement—namely, that the material should have a high refractive index if its nonlinearity is to be sufficient.

At the same time, efforts to investigate the nonlinear properties of niobates were in progress at the Bell Laboratories. The first major advance in the development of nonlinear materials came as a result of this effort, with the announcement of the properties of lithium meta–niobate (LiNbO$_3$).[26,117] Lithium niobate, as it is normally called, offered several attractive features over the then popular materials, ADP and KDP. It is nonhygroscopic and hard, taking a good polish readily. After some initial problems, the substance proved to be mechanically stable, not unduly affected by thermal or mechanical shock; most important, it had a large nonlinear coefficient relative to KDP. Moreover, the temperature sensitivity[117] of its birefringence was such that, by varying the temperature, phase matching could be achieved in the x–y plane of the crystal (i.e., at 90° to the optical axis).

One puzzling property of lithium niobate, however, is that the precise refractive indices seem to vary from boule to boule. As we shall see shortly, from this it became clear that in order to be acceptable the optical quality of a nonlinear material has to be much better than it would have to be for a normal linear optics application.

For a brief period, it looked as if lithium niobate might be an almost ideal material for nonlinear interactions in the visible and near-infrared regions. However, when really good quality crystals became available it was discovered that the crystal suffered from a hitherto unknown damage effect.[7] In other materials, mechanical damage had been seen at very high power levels (of the order of megawatts per square centimeter), but lithium niobate showed a damage effect when illuminated by a continuous-wave gas laser. The effect was manifest as a track of slightly altered refractive index in the crystal, following the path of the laser beam. For many practical purposes it might have been overlooked, as it had been in the past; but for efficient phase-matched interactions in which the relative phase of the waves is important, the effect was devastating. On the other hand, the phenomenon has been put to use for such applications as holography.

Many theories were advanced to explain this damage effect.[42] Although it was discovered rather early that the effect could be ameliorated to some degree by heating the crystal, it has been shown only recently that a major cause of the effect is an iron impurity in the material.[135] Despite this damage effect, and perhaps because good quality, large samples of the subsequently discovered crystal barium sodium niobate[143] were hard to grow, lithium niobate remains a very important nonlinear material. The first operative pulsed and continuous-wave parametric oscillators were made with it, and the first engineered parametric oscillator offered for sale used a crystal of lithium niobate. The properties of this material are discussed in more detail later.

Following lithium niobate, a host of related ferroelectric niobates were grown and studied. Of these, the most significant to date has proved to be barium sodium niobate, although many of the others have their own particular features. Comprehensive tables of materials properties are given in Appendix 2.

While this work was proceeding, a great deal of research was done to define the characteristics of a "good" as opposed to a "bad" sample of a given crystal, since the criterion was not just simple optical homogeneity in the sense that this quality is normally measured in an interferometer.

Much work was also carried out in the search for a theory that would allow the precise calculation of the refractive index and of Miller's Δ of a given material. In 1969 Levine,[99] starting from theoretical work by Phillips and Van Vechten,[159] ascribed the nonlinear terms to the anharmonic motion of bond charges located approximately half-way between the neighboring atoms. Using this model, he was able to make accurate calculations of the second- and third-order nonlinear susceptibilities of semiconductors with the zincblende structure and of materials with the wurtzite-, the diamond-, and the rock salt structure and of the second-order nonlinear susceptibility of alpha quartz. By a subsequent modification of this model, Levine[100] was able to explain the negative sign of the nonlinear susceptibility of ZnO observed by Miller and Nordland.[120]

Aided by this theory, Chemla[41] accurately calculated the refractive indices and Miller's Δ of several compounds, using only the crystallographic parameters and the covalent radii of Phillips and Van Vechten.

An anomalously small value of Miller's Δ, occurring in beryllium oxide,[87] was explained both in terms of Levine's theory and in terms of a different theory developed by Jerphagnon.[88]

These results spurred the search for new materials that would have the high nonlinear susceptibility of gallium arsenide but would, in addition, be birefringent. As a glance at Table A2.2b shows, such materials are being found in the chalcopyrite structures belonging to the crystal class $\bar{4}2m$.

One other technique that must be mentioned in this brief review, before we begin to discuss details, is the Kurtz powder technique. In 1968 S. K. Kurtz[95] described an important new technique for making a quick survey of many materials. Until this time, it had been necessary to obtain for study a single crystal of a material before its nonlinear coefficients could be measured and before it could be stated that phase matching was or was not possible.

Since the development of a technique to grow even a small single crystal of a new material is a time-consuming process, the search for new materials had been extremely slow and unrewarding. Kurtz described a method by which a good estimate of the nonlinearity could be obtained from a crystalline powder alone, using particles typically of 10-μ size. The method also made it possible to determine whether phase matching could occur. This approach allowed Kurtz and others to survey quickly a huge number of materials and to spot some promising candidates for crystal growing.

In this chapter we cover the techniques for measuring the nonlinear coefficients of powders and single crystal specimens. We also discuss the method for telling a good sample from a bad one of the same material. Finally, we present in some detail the properties of the currently popular nonlinear materials, lithium niobate, barium-sodium niobate, lithium iodate, ADP and KDP, and proustite.

4.2 QUALITY ASSESSMENT OF NONLINEAR MATERIALS

We have already remarked that the normal assessment of optical quality, such as is made on a Twyman–Green interferometer, is not a sufficient test for nonlinear materials. As our basic premise, then, we assume that a non-linear material to be used in any device application will be of excellent linear optical quality, typically having less than one fringe of optical distortion per centimeter of aperture and of length. We also assume that the material will be essentially free from any absorption at the frequencies at which we wish to operate. With these requirements as a starting point, we now investigate the other qualities that the nonlinear crystal must possess.

The most important additional quality requirement is easily understood, and it follows directly from an understanding of the phase-matching process. The nonlinear interaction—be it SHG or parametric gain, oscillation, or up-conversion—is a traveling-wave process in which the direction of energy transfer is controlled by the instantaneous relative phases of the waves concerned. Phase matching involves the careful adjustment of the indices of refraction for the different frequencies and polarizations being matched, so that, if the correct phase relationship exists at the entrance to the crystal, it will remain correct through the crystal.

A change in index at any point in the crystal such that one wave relative to the others shifts phase by π will reverse the direction of energy flow between the waves. Thus, in SHG, the starting phases automatically adjust themselves to favor growth of the second harmonic at the expense of the fundamental. However, if at some point in the crystal the extraordinary or the ordinary index of refraction varies slightly from perfect phase matching, this phase relationship is destroyed and a second wave of SHG, out of phase with the first one, is generated. As a result, the SHG already generated starts to be attenuated. How can we detect such effects?

Examination of a crystal in a Twyman–Green interferometer shows the overall optical length, that is to say

$$l_{\text{opt}} = \int_0^L n(x) \, dx$$

Here we imply by $n(x)$ the refractive index as a function of x, the distance measured through the crystal in the direction of observation. If we examine a uniaxial crystal along any observation direction, but perpendicular to the symmetry axis (c axis), we find that l_{opt} varies with the polarization of the light used. Type I phase matching requires that, for SHG, the indices for the second-harmonic wave of one polarization and the fundamental wave of the orthogonal polarization be equal. Thus at least for the phase-matching direction, we would expect to find that

$$l^o_{opt}(\omega) = \int_0^L n^o(x, \omega)\, dx \qquad (4.1)$$

$$l^e_{opt}(2\omega) = \int_0^L n^e(x, 2\omega)\, dx \qquad (4.2)$$

$$l^o_{opt}(\omega) = l^e_{opt}(2\omega) \qquad (4.3)$$

However, the condition for perfect phase matching is more stringent than this, since the relative phases are not compared only by the traveling-wave interaction at the entrance and the exit of the crystal, but at each point in between. Thus we require, in addition to equations 4.1 to 4.3, the following condition:

$$\int_0^{L'} n^o(x, \omega)\, dx = \int_0^{L'} n^e(x, 2\omega)\, dx \qquad \text{for} \quad 0 \le L' \le L \qquad (4.4)$$

Equation 4.4 seems both obvious and trivial, but in practice it is neither. This is because in most if not all of the nonlinear materials of interest, and particularly in the ferroelectric niobates, the refractive indices are sensitively dependent on the chemical composition of the crystal. Again, this does not sound serious until it is realized that the very process of crystal growing tends to alter this chemical composition. For example, lithium niobate is grown by the Czochralski method from a melt composed of Li_2O and Nb_2O_5 mixed in stoichiometric or other quantities. "Other quantities" are chosen because, when a crystal is pulled from a stoichiometric melt of $LiNbO_3$, the grown crystal is not exactly $LiNbO_3$ but differs by a small amount from the exact composition. Thus we find that as the crystal is pulled, if we start with a stoichiometric melt, we finish with a melt that is rich in Nb_2O_5 and a crystal that starts rich in Li_2O, but becomes less so as it grows. It is most important to note that the crystal composition varies along its length as it grows; also, it can vary across its section, since local fluctuations in temperature can perturb the growth. Thus a typical crystal is not of constant composition, and the refractive indices can vary from point to point throughout its volume. The Twyman–Green interferogram of a poor quality lithium niobate crystal

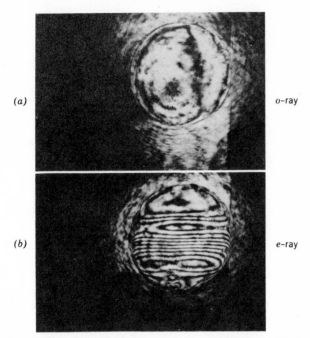

(a)   o-ray

(b)   e-ray

Figure 4.1 Twyman–Green interferograms for a crystal of lithium niobate showing the effect of composition variations: (a) Ordinary ray, (b) extraordinary ray.

(Figure 4.1) illustrates the effect well. The interferogram is shown for ordinary and extraordinary rays, along the a or x axis of the crystal. Notice that the two are quite different, although each on its own exhibits a reasonably uniform crystal. Thus this crystal would not even meet our first condition for phase matching (equation 4.3).

In order to test for the validity of the second condition (equation 4.4), further interferometry could have been done, once a crystal had been obtained that looked perfectly uniform and similar to both polarizations of light. However, the established and most sure means for checking a crystal is to measure SHG in it. For the niobates in which the phase matching is temperature sensitive, the usual apparatus is sketched in Figure 4.2. A helium–neon laser (1.15 or 1.08 μ) or a Nd:YAG (neodymium–yttrium-aluminum-garnet) laser operating at 1.06 μ is used to provide the source, and the SHG is monitored on a pen recorder as the temperature of the crystal is tuned through the phase-matching temperature. Angle tuning can be used with equal success. A perfect crystal gives a plot of second-harmonic power versus temperature that is essentially a $[(\sin x)/x]^2$ function (see equation 2.44 and Figure 3.2. If various portions of the crystal phase match at different

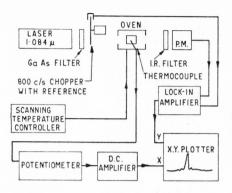

Figure 4.2 Apparatus for measuring the SHG response of a crystal using temperature tuning. For angle tuning, the oven and controller would be replaced by a turntable with angle pickoff to the $x-y$ plotter. After Midwinter, *J. Appl. Phys.*, **39**, 3033 (1968).

temperatures because of varying compositions, the central peak of the SHG is broadened and reduced in height, and the side lobe pattern is distorted close to the center. Further out, little distortion is detected, since the SHG occurs from ever-decreasing lengths of the crystal as the phase mismatch increases and the index error in the small length becomes negligible.

Thus the standard acceptance test has become to measure the profile of the SHG as a function of temperature[38] or angle as the crystal is tuned through the phase-matching condition. It should be noted, however, that this is not a completely sufficient test, since under the right conditions small but oscillating variations in the relative indices[122,150] (i.e., in the composition) can lead to response curves that look like $[(\sin x)/x]^2$ functions but have a reduced peak amplitude, with the energy lost from the central peak appearing in anomalously large side lobes. Thus it is important to examine the ratio between the central peak and the first side lobe, to make sure that this ratio has the value given by the appropriate function.

The numerical quotation for quality of a nonlinear crystal is usually given in terms of an effective interaction length, which is less than or equal to the length of the crystal in the direction of the traveling-wave interaction.[38,111] The value of this effective length is arrived at in the following manner. The second-harmonic power in an ideal crystal varies as $\sin^2 (\frac{1}{2}\Delta k \cdot L)/(\frac{1}{2}\Delta k \cdot L)^2$ where Δk is the phase mismatch, typically as a function of temperature or angle, and L is the length of crystal. If we confine our attention to a crystal like lithium niobate in which temperature tuning of the refractive indices is used to achieve phase matching, then we can define a temperature interval ΔT which is the temperature interval between the half-power points of the curve of second-harmonic power against temperature. This can be related to the crystal properties by the following equation:[38]

$$\Delta T = 0.89\lambda(2\omega)\bigg/ L\left(\frac{\partial n^{\mathrm{o}}(\omega)}{\partial T} - \frac{n^{\mathrm{e}}(2\omega)}{\partial T}\right)_{T_{\mathrm{pm}}} \qquad (4.5)$$

This relation defines the theoretical temperature width for a perfect crystal. The experimentally measured temperature width is related to an effective interaction length l_{eff} by the following relation, which serves as a definition of l_{eff}:

$$\Delta T_{exp} = 0.89\lambda(2\omega) \bigg/ l_{eff}\left(\frac{\partial n^o(\omega)}{\partial T} - \frac{\partial n^e(2\omega)}{\partial T}\right)_{T_{pm}} \tag{4.6}$$

Hence, once the value of the factor $(L \cdot \Delta T)$ has been calculated from equation 4.5, the effective length l_{eff} is obtained directly from the experimentally measured half-power temperature width by the relation

$$l_{eff} = \frac{\Delta T \cdot L}{\Delta T_{exp}} \tag{4.7}$$

The effective length so obtained is a good measure of the active length of the nonlinear crystal or the length of crystal over which phase matching is maintained. If the effective length is equal to the length of the crystal under study, then that crystal is of very good quality. Notice that the theoretical value of the factor $(L \cdot \Delta T)$ is constant for a given material and interaction (set of frequencies), but it varies for each new material. It also varies for different sets of operating frequencies in the same material. The foregoing results can be recast for angle tuning using differentials with respect to angle to define again an effective interaction length which is a measure of crystal quality.

4.3 THE ACCURATE MEASUREMENT OF OPTICAL NONLINEARITY

A detailed study of the published measurements of optical nonlinearities quickly shows that the measurements are seldom very precise and that they are nearly always comparative rather than absolute determinations. The reasons for this are easy to understand. To make an absolute determination, it is necessary to measure with high precision the fundamental power, the second-harmonic power, and the spatial and temporal mode structure of the laser beam used. All these measurements are difficult to make with certainty, and all are crucially important in an absolute determination. (The reason for the temporal mode distribution will become clear in Section 5.4.)

The most widely employed and best tabulated reference material is ADP. The optical nonlinearity of this material was measured with great care by François[54] in 1966, using an unfocused beam from a 6328-Å helium–neon laser operating in single transverse mode and single or multiple longitudinal mode. He found a value of $(1.36 \pm 12\%) \times 10^{-9}$ esu. This measurement was further strengthened in 1967 by a remeasurement of ADP, carried out by Bjorkholm and Siegman,[17] using a focused 6328-Å helium–neon laser. They concluded that the optical nonlinearity of ADP was $(1.38 \pm 16\%) \times 10^{-9}$ esu. A reading of the original papers gives a good idea of the care necessary in making such a measurement.

Although ADP is well characterized, other materials are also used as references in the published literature. Particularly common in this respect are quartz and potassium dihydrogen phosphate (KDP). A very careful comparison of the three materials—quartz, KDP, and ADP—was made by Jerphagnon and Kurtz,[85] and now it is possible to compare measurements made with reference to any of these.

How is an actual measurement made? In the case of the absolute determination, the formula relating SHG to fundamental power (equation 2.52) is evaluated very carefully using the experimental data for beam profile, crystal length, and orientation in the phase-matching direction. The laser frequency output is measured to discover the number of oscillating modes and the power distribution, and the second-harmonic power generated is measured as the crystal is tuned through phase matching. It is then possible to derive a number for the optical nonlinearity. However, great care must be taken if the measurement is to be meaningful.

For example, it is easy for a plane-parallel crystal to act as a resonator for the fundamental frequency and, hence, to enhance the SHG. This problem has been neatly overcome by Wynne and Bloembergen,[168] whose technique deliberately uses a wedge-shaped sample and tunes through coherence lengths by sliding the wedge through the laser beam. If focused beams are used, accurate allowance must be made for the effects of beam walk-off between ordinary and extraordinary rays, and when multiple longitudinal modes are present, allowance must be made for the enhancement due to them.

A comparative measurement is easier. For one thing, it is not necessary to make absolute measurements of optical power. However, more than that, the customary technique does not rely on phase-matched generation. For this reason the quality of the crystal used is less crucial and the characteristics of the laser need not be known precisely, since the measurement in the reference sample is equally affected. The method is known as the Maker fringe technique, after the first report of its use by Maker et al.[105] in 1962. A plane-parallel slab of crystal is oriented so that the coefficient to be studied will be dominant in the interaction. For example, to study $d_{36} = d_{zxy}$ in KDP, a slab should be cut with its faces containing the z axis and the face normal at 45° to x and y. A plane-polarized laser beam entering normally, polarized at 90° to z, would then have equal components of E_x and E_y. The SHG produced would be polarized along z orthogonally to the input. But the interaction in this geometry is not phase matched, and a small signal results. As the crystal is rotated by small amounts in the plane of the z axis and the beam direction, the SHG fluctuates as the effective length of the crystal and the phase mismatch vary. The effect is to produce "fringes" in the SHG output which are really caused by operating the SHG at large Δk. As shown in Chapter 3 the SHG varies as $\sin^2 (\Delta k \cdot L/2)$. Suppose that, for the direction of propagation, the refractive indices for the second-harmonic and fundamental beams are $n(2\omega)$ and $n(\omega)$, respectively; then the phase mismatch is given by $\Delta k = (n(2\omega) - n(\omega)) \, 2\omega/c$. If the common internal **k** vector direction for the SHG and fundamental is at θ to the face normal, then the effective thickness of the material is $L \cos \theta$, and we have our result that the SHG power varies as

$$W(2\omega) \sim \sin^2 \left[\frac{[n(2\omega) - n(\omega)]L\cos\theta}{c} \right] \tag{4.8}$$

The Maker fringes stem from the variation of θ over fairly small angles, typically 5 to 15°. By measuring the separation of the Maker fringes, a value for $[n(2\omega) - n(\omega)]$ can be determined. This in turn leads to a value for the effective length of material generating the observed SHG. Then by extrapolating the peak SHG in the $\theta = 0$ direction and comparing it with the value obtained for a reference sample, making due allowance for the different effective lengths, a comparative value for the nonlinear coefficients is obtained. The form of the Maker fringes was illustrated in Figure 3.2.

The foregoing discussion of the Maker fringe method was simplified somewhat, but the basic techniques were presented. A much more detailed and critical appraisal of the technique was given by Jerphagnon and Kurtz,[85] who assessed the importance of corrections in the measurements for multiple reflections within the sample, finite beam size, absorption, and optical activity. They concluded that, with care, the technique can yield measurements that are accurate to 5%. This exceeds the accuracy of the fundamental determination of the second-harmonic coefficients in ADP.

The Maker fringe technique has been extended by several workers to the measurement of the relative sign of the nonlinear coefficients between two crystals.[120,168] The technique uses two crystals in series in the same laser beam so that interference effects are observed in the SHG between them. The crystals are mounted in an evacuated chamber which is slowly filled with a known gas, such as carbon dioxide. The dispersion in the gas path between the two crystals causes the relative phases of the fundamental and the second harmonic from the first crystal entering the second to vary, and interference effects are observed in the net SHG. By carefully measuring these effects, the required result is obtained.

Relative measurements of the optical nonlinearity may also be made using phase-matched interactions. In this case, the laser beam is split so that two crystals are used simultaneously as the reference and the sample. Each is phase matched, and the relative second-harmonic powers are monitored. However, it remains necessary to treat the raw experimental data with great caution before drawing conclusions about the values of the coefficients. An excellent discussion of the corrections needed in this type of measurement has been given by Nash et al.[121] in the course of a detailed study of lithium iodate.

It should be noted that all the foregoing techniques require good crystals of the materials to be studied. It is not necessary to have the highest quality, but usually clear uniform optical quality material is mandatory. In the next section we describe the Kurtz powder technique, which eliminates this requirement.

4.4 KURTZ POWDER ASSESSMENT OF NONLINEAR MATERIALS

The Kurtz powder technique represents the first real means of screening, experimentally, large numbers of unknown materials for nonlinear activity, without having to perform the slow and expensive task of growing good quality crystals of each material. Kurtz[95,97] showed that it is possible, by measurements on powders, to ascertain whether a crystal has large or small nonlinearity and whether it can be phase matched. Given this information and the commonly known data on crystal properties, the probability of selecting for crystal growth a material that will subsequently be useful is drastically increased.

The apparatus used is illustrated in simplified form in Figure 4.3. A laser, typically a Q-switched high-repetition rate Nd:YAG laser, serves to illuminate a large area of a thin layer of powder of the material to be studied. The

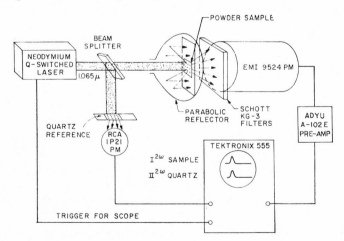

Figure 4.3 Schematic layout of the apparatus for use in the Kurtz powder measurement. After S. K. Kurtz, *IEEE J. Quant. Electr.*, **QE-4**, 578 (1968).

powder is compacted into a thin cell to define the thickness of the layer. The cell is held within an optical system that collects all the scattered SHG from the sample over 4π radians, and the signal is monitored by a photomultiplier. The measurement technique is to study the amount of SHG from the powder sample as a function of particle size in the sample, the powder being sieved to a small range of size for each test. It is necessary to monitor the laser power for control purposes. The required information is obtained from the measurement of SHG power versus particle size, which is compared, for an identical geometry, with the results determined for quartz powder samples. This is possible because of two distinctly different responses by phase-matchable and non-phase-matchable materials.

For very small particle sizes, such that $r \ll l_{coh}$* for any arbitrary orientation, the total integrated SHG varies as the particle radius r. The effect comes about because the interaction is effectively always phase matched, since the radius is not great enough for any serious phase error to occur. Thus, as the radius increases, the interaction efficiency increases rapidly, but it is offset to some extent by the decrease in the number of particles.

For a non-phase-matchable material, once the particle size passes the average coherence length l_{coh} (so that for most or all orientations phase-mismatch effects become apparent), the SHG varies inversely as the particle size because the SHG per particle does not increase as rapidly as the number of particles decreases. Thus a plot of SHG against particle size yields a curve of the form shown in Figure 4.4, with a pronounced maximum.

However, for a material that is phase matchable, once the particle size reaches the average coherence length (averaged for all crystal directions, not

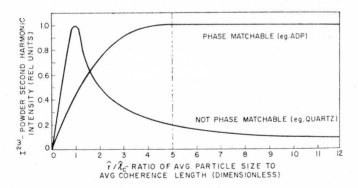

Figure 4.4 The typical response for powders of phase-matchable and non-phase-matchable crystals, showing the SHG as a function of particle size. After S. K. Kurtz, *IEEE J. Quant. Electr.*, **QE-4**, 578 (1968).

* $l_{coh} = \pi c / \{[n(2\omega) - n(\omega)]\omega\}$

just the phase-matching direction) the gain in SHG from the particles that are correctly oriented approximately balances the loss in SHG from the decrease in the number of particles. The net result is that the overall SHG remains essentially constant, as shown in the second curve of Figure 4.4. Thus, by plotting the SHG as a function of particle size and by inspection only, it is possible to decide whether a material allows phase matching.

By the use of this technique, Kurtz surveyed a very large number of compounds of potential interest and classified them so that each falls into one of five groups: large coefficient, small coefficient, phase matchable, not phase matchable, and centrosymmetric materials.

4.5 LITHIUM NIOBATE

The uniaxial crystal class of lithium niobate is rhombohedral R3c, trigonal point group 3m. As such, four nonzero coefficients—d_{22}, d_{15}, d_{32}, and d_{33}—are allowed for the nonlinear polarizability, although very likely $d_{32} = d_{15}$ because of the Kleinman symmetry conjecture. The measured values of these coefficients are given in Appendix 2.

Extensive studies of the growth and crystal structure of lithium niobate have been given in the literature. The most complete group of papers, published in 1966,[123] describes growth, domain structure, and crystal structure studies. Subsequent papers have highlighted some of the practical problems that arise in the use of lithium niobate.

Generally speaking, lithium niobate crystals are pulled with the c (z) axis as the pulling direction, although a (x) is also used. The former crystals can be made single domain during the pulling process by passing a small poling current through the pull rod. An a-axis crystal must be poled in a separate operation after growth. The action of poling is to align all the ferroelectric domains in the same direction; it is achieved by holding the crystal at a sufficiently high temperature for the domain direction to be reversible and then aligning all the domains in a single direction with an electric field. Subsequently the crystal is cooled until the domains are no longer reversible, and the field is removed. The probable action of poling at the atomic level has been discussed by Niizeki et al.[129] The crystal structure of lithium niobate is illustrated in Figure 4.5, which indicates that the lithium, niobium, and oxygen atoms lie in layers. The domain direction, like the positive z direction, is determined by the position of the layer of lithium atoms in the structure. Poling causes this layer of lithium atoms to move through the oxygen layer into the adjacent space, reversing the sense of direction of the crystal structure. This can occur above about 1150°C, and Niizeki has shown that at about this temperature the crystal structure has expanded sufficiently for the oxygen layer to allow the passage of the lithium atoms without severe

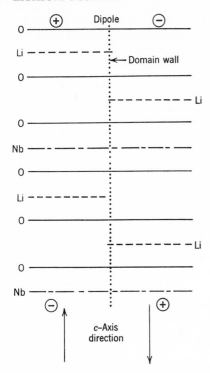

Figure 4.5 Schematic diagram of the structure of lithium niobate showing the effects of poling. The layers represent layers in the crystal of lithium, niobium, and oxygen atoms.

distortion. Below this temperature, the structure contracts and the lithium atoms are frozen in position. Since the crystal melts at about 1250°C, care must be taken not to melt the crystal if the poling operation is done after growth.

The importance of poling can be seen from a study of the expression for the nonlinear coupling for SHG in lithium niobate (see Table 3.2). For two ordinary rays mixing to generate an extraordinary ray, the effective non-linearity is given by

$$d_{\text{eff}} = d_{15} \sin \theta + d_{22} \cos \theta \sin 3\varphi$$

Thus, for propagation in the x-y plane, $\theta = \pi/2$ and $d_{\text{eff}} = d_{15} = d_{xxz}$. Changing the direction of z or the direction of the domain alignment, there-fore, changes the sign of the nonlinear coupling and reverses the direction of energy flow. Thus, if the crystal is multidomain, the z direction alternates at somewhat random intervals and the direction of energy flow is not con-tinuous but alternating. A secondary effect is that the domain wall region in the crystal (i.e., the boundary region between two antiparallel domains), shows up optically as a region of strain or refractive index perturbation

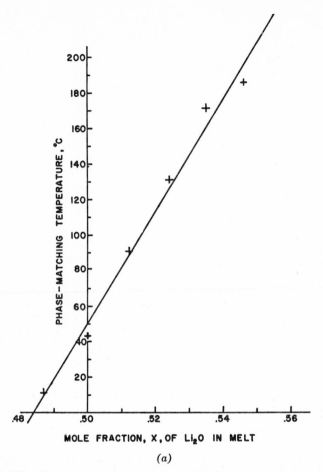

Figure 4.6 The effects of composition on phase-match temperature for lithium niobate: (a) for 1.06 μ (neodymium) after Fay et al., *Appl. Phys. Lett.*, **12**, 89 (1968); (b) for 1.15 and 1.08 μ (helium–neon), after Bergman et al., *Appl. Phys. Lett.*, **12**, 92, (1968).

because of the distorted internal fields. This leads to heavy scattering loss and distortion of the phase front of a traveling wave.

Once the crystal is poled, it is possible to assess its linear and nonlinear quality in the way that we have already described, although some fabrication must be carried out to produce a pair of good quality flat faces for a laser beam to enter and the second-harmonic beam to leave. The usual procedure with lithium niobate is to heat the crystal to achieve the phase-matching temperature for the particular laser source being used and then to plot

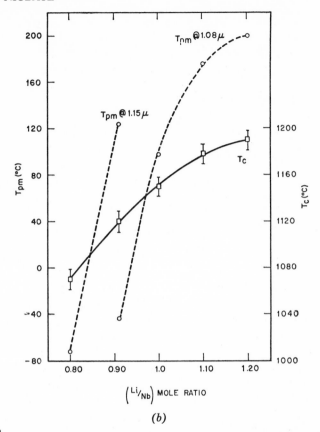

Figure 4.6b

the SHG against temperature. Figure 4.6 plots phase-matching temperature against composition of melt for lithium niobate. Notice that the phase-matching temperature can vary by huge amounts for different compositions. Thus, if nothing is known of the composition of a particular crystal, a wide range must be searched, and it may be necessary to use more than one laser line.

To predict the phase-matching temperature for other than SHG requires data on the refractive indices at all the frequencies concerned, as a function of both composition and temperature. In general, this information is not available, but it can be approximated by the use of the data that are at hand. Hobden and Warner,[77] in a careful study of the refractive index of lithium niobate grown from a stoichiometric melt, concluded that the refractive-index data for all temperatures and wavelengths of practical interest could be summarized in two Sellmeier equations, one for the extraordinary index and one for the ordinary. These equations are ideally suited to use in a digital computer, and since most phase-matching calculations are done using a computer, the equations have been used extensively. Point-by-point data have also been published. The equations derived by Hobden and Warner are, for the ordinary ray:

$$n_o{}^2 = 4.9130 + \frac{1.173 \times 10^5 + 1.65 \times 10^{-2}T^2}{\lambda^2 - (2.12 \times 10^2 + 2.7 \times 10^{-5}T^2)^2} - 2.78 \times 10^{-8}\lambda^2$$

and for the extraordinary ray:

$$n_e{}^2 = 4.5567 + 2.605 \times 10^{-7}T^2$$

$$+ \frac{0.970 \times 10^5 + 2.70 \times 10^{-2}T^2}{\lambda^2 - (2.01 \times 10^2 + 5.4 \times 10^{-5}T^2)^2} - 2.24 \times 10^{-8}\lambda^2$$

where the temperature T is in degrees Kelvin and the wavelength is in nanometers (10^{-9} m; 1 μ = 10,000 Å = 1000 nm).

Using these equations, it is now a simple matter to derive a value for the refractive index at any temperature or wavelength throughout the visible or near-infrared range and at temperatures up to several hundreds of degrees centigrade. The authors estimate that their data are good for use from 4000 Å to a least 4 μ, and for temperatures from around 0°C to over 400°C.

In order to adjust the predictions of phase-matching temperature made from the foregoing equations for stoichiometric material to a prediction for some other but known composition, it is generally a good approximation to add to the predicted temperature the value of [T_{pm} (new composition) − T_{pm} (stoichiometric)] taken from the published SHG data.[13,51] The approximation involved in this operation is the assumption that the form of the index change produced by composition is the same as that produced by temperature. Thus, if the temperature changes the refractive index uniformly at all wavelengths but leaves the dispersion unchanged, we are assuming that the composition has a similar effect. Although this is not quite true, it is in general an adequate approximation.[112a]

The importance of the composition dependence of the phase-matching temperature in lithium niobate is twofold. Clearly, an investigator must know what to expect before designing an experiment, so that a crystal oven can be given the right operating range. But the second and much more important reason concerns damage. We have already mentioned that when lithium niobate is illuminated at room temperature with a low-power continuous-wave gas laser, a track appears through the crystal of different refractive index.[7] The change in birefringence is typically 0.0001 (i.e., [n_e − n_0] is typically -0.0001).[42] In some material, this damage starts to fade slowly as soon as the laser beam is removed, typically with a half-hour time constant, but in other material the track can be quite stable for days. However, in all crystals of lithium niobate, the damage anneals out almost instantaneously if the crystal is heated to about 180°C or more. Thus, if the phase-matching temperature for the interaction of interest is in this temperature range, it is unlikely that refractive-index damage will be a serious problem.

Hence there is a great interest in finding a means of modifying the crystal so that a given interaction, involving visible radiation that is capable of damaging the room temperature crystal, shall be phase-matched at a high temperature. The early reports of the composition dependence of the phase-matching temperature apparently offered such a solution. This approach, however, introduces other problems that are due to the composition-dependent nature of the crystal quality. In general, only one uniform growth composition is allowed by the Li_2O/Nb_2O_5 phase diagram.[38] The particular mixture that yields uniform composition crystals by Czochraski pulling, the congruent melting one, is $Li_2O = 48.6\%$ and $Nb_2O_5 = 51.4\%$ by weight. Crystals of this composition have been grown, and their uniformity properties were essentially ideal. However, for frequency doubling of the Nd:YAG laser, this material is not ideal because its phase-matching temperature occurs at -8°C. Cooling, which always is less convenient than heating, is involved at this temperature, and serious index damage would occur as well.

The other important application for lithium niobate is in parametric oscillation, and for an oscillator pumped in the green, much of the useful tuning range occurs below the anneal temperature for index damage. Hence a search has been made for other solutions to the problem.

Growing material of noncongruent melting composition has resulted in poor crystal quality and in spatially varying phase-matching temperatures, and both conditions are ruinous for nonlinear optical interactions. A more promising solution has been proposed by Bridenbaugh et al.,[34] who showed that it is possible to dope the crystal with magnesium oxide. They added 1.0% MgO by weight to the melt, without apparently altering the quality of the crystal for nonlinear optical interactions; in so doing, however, they succeeded in raising the phase-matching temperature by some 50°C. It remains to be seen how far this technique can be extended.

Thus we are left with a crystal that, after a great deal of study, is found to have many attractive features. As such, it has been featured in many of the major advances in experimental nonlinear optics, but several unresolved problems remain regarding its use in applications other than specialized interactions.

4.6 BARIUM SODIUM NIOBATE

Barium sodium niobate[143] is another ferroelectric material, one of the many that were studied following the initial success of lithium niobate. The crystal is biaxial, although not greatly so, in that the refractive indices for light polarized along the three principal axes are related as follows: $n_z \leqslant n_y \simeq n_x$; $n_y - n_z = 0.12$ and $n_x - n_y = 0.002$ at room temperature. To a first approximation, then, the material is uniaxial, at least insofar as the gross properties of phase matching are concerned. The crystal has the formula $Ba_2NaNb_5O_{15}$, and there are two formula units per unit cell. The similarity to lithium niobate can be seen when the latter is written as $Li_5Nb_5O_{15}$. The point group below about 300°C is orthorhombic mm2. It is similar to $LiNbO_3$ in many of its device properties.

The major reason for the prominence of barium sodium niobate as a nonlinear optical material is that, above room temperature, it does not suffer from the optical damage effects that have so limited the applicability of lithium niobate. This property is believed to be attributable to the structure, which is filled—a characteristic shared with other niobates having the filled tungsten bronze structure. This freedom from damage has made possible very efficient SHG from a continuous-wave Nd:YAG laser, providing an intense source of 5300-Å radiation. It has also been exploited in the building of a continuous-wave parametric oscillator having the remarkably low threshold for oscillation of 3 mW of pump power at 5300 Å.

However, the use of barium sodium niobate presents other problems, stemming from its fundamental properties (some of them differing from those of lithium niobate). The crystals are normally grown from a melt by the Czochralski method at about 1440°C. The first problem encountered has been one of striations in the material, which have been attributed to growth rate or temperature fluctuations at the melt interface. Careful control of the growth conditions has largely eliminated this problem and has yielded material that is more uniform in composition and optical properties. There is also a problem similar to that already discussed in connection with lithium niobate—namely, that of compositional fluctuations being reflected as refractive-index variations or striations.

The as-grown crystal is not ferroelectric and only becomes so below the Curie temperature of approximately 585°C. Below this temperature, the crystal shows the presence of multiple domains which, as with lithium niobate, correspond to regions with alternately oriented c axes. These can be removed by poling the crystal through the application of an electric field along the c-axis direction, with the crystal held above the Curie temperature. The details of poling this material have been given by Singh et al.[143] Basically, the procedure consists of pasing a current not exceeding 0.5 mA through the crystal, which is held at 650°C in an oxygen environment. The material is then slowly cooled to room temperature, maintaining the electric field; after cooling it is stable with regard to domains.

However, a new problem occurs in barium sodium niobate when, at a temperature of about 300°C, the crystal undergoes another phase transition, from tetragonal to orthorhombic. Unless precautions are taken, the material suffers from micro twinning, after undergoing this transition. That is to say, small areas of the crystal alternate in their a and b axis directions, relative to some fixed external coordinates, but they generally retain the c axis alignment imparted to them by the poling process. The micro twinning can be removed by heating the crystal to a temperature above the 300°C phase transition but well below the Curie temperature (585°C) and then cooling the crystal slowly with a compressive stress of some 10^7 dynes/cm² applied along one of the axes other than the c axis. Below the phase-transition temperature, the axis to which the compressive stress was applied becomes the a, or x axis. In commercially available material, both the detwinning and poling are normally undertaken by the supplier as a matter of course.

The optical transmission range of barium sodium niobate extends from the ultraviolet absorption edge at about 3700 Å through to about 5 μ, with a small absorption near 3 μ attributed to hydroxyl group contamination. The same absorption has also been noted in lithium niobate. Barium sodium niobate from some sources also shows a marked brown discoloration. This is reputed to occur only in material from iridium crucibles, the niobate grown from platinum crucibles being clear.

The refractive indices have been measured in detail at room temperature and are presented in Table 4.1 and 4.2. They are not available as a function of

TABLE 4.1 REFRACTIVE INDICES OF BARIUM SODIUM NIOBATE AT ROOM TEMPERATURE.*

λ (nm)	n_x	n_y	n_z
457.9	2.4284	2.4266	2.2931
476.5	2.4094	2.4076	2.2799
488.0	2.3991	2.3974	2.2727
496.5	2.3920	2.3903	2.2678
501.7	2.3879	2.3862	2.2649
514.5	2.3786	2.3767	2.2583
532.1	2.3672	2.3655	2.2502
632.8	2.3222	2.3205	2.2177
1064.2	2.2580	2.2567	2.1700

* After Singh et al., *Phys. Rev.* **B2**, 2709 (1970).

TABLE 4.2 BARIUM SODIUM NIOBATE CONSTANTS
FOR USE AT ROOM TEMPERATURE IN A SINGLE
TERM SELLMEIER EQUATION OF THE FORM*

$$n^2 - 1 = S_0 \lambda^2 / (\lambda^2 - \lambda_0^2)$$

Refractive index	S_0	λ_0 (nm)
n_x	3.9495	200.97
n_y	3.9495	200.35
n_z	3.6008	179.44

* After Singh et al., *Phys. Rev.* **B2**, 2709 (1970).

temperature in the very convenient form given for lithium niobate. However, partial data have been published, and these are summarized in Figure 4.7. Phase matching for the SHG of 1.06-μ radiation from the Nd:YAG laser occurs at about 84 and 96°C for propagation along the x and y axes, respectively. As with lithium niobate, however, departures from stoichiometry of the crystal composition can change these figures; typical ranges are 80 to 100°C, and 90 to 110°C, respectively.[143]

The nonlinear coefficients for barium sodium niobate have also been measured in detail. There are five independent coefficients. They are d_{31}, d_{32}, d_{33}, d_{24}, and d_{15}. Of these, the Kleinman condition predicts that $d_{31} = d_{15}$ and $d_{32} = d_{24}$ and this has been found to be true to within the accuracy of measurement. The values obtained are listed in Appendix 2. It is apparent from these results that barium sodium niobate offers another advantage over lithium niobate—that of larger nonlinear coefficients. Thus, in summary, barium sodium niobate is more nonlinear than lithium niobate and does not suffer from index damage. Against this must be set the greater difficulty that has been experienced in growing high optical quality crystals, and the problems that are met in the poling and detwinning procedures. At this time, it appears that each material has its advantges, depending on the particular application.

4.2 ADP AND KDP

ADP and KDP were the first crystals used for the demonstration of phase-matched SHG. Since then they have served in many other interactions. They are also widely used as electrooptic materials. Both are negative uniaxial crystals with a transmission band from the ultraviolet to the near infrared, typically 2000 Å to 1.5 μ. They belong to point group $\bar{4}$2m and, thus, have tetragonal symmetry.

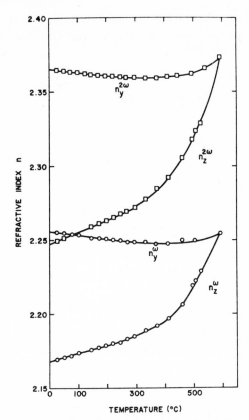

Figure 4.7 The temperature dependence of the refractive indices of barium sodium niobate at 1.064 and 0.532 μ. After Singh et al., *Phys. Rev.* **B2**, 2709 (1970).

Both crystals were first developed as piezoelectric materials and have been used extensively for ultrasonic transducers. As a result, large crystals of ADP—approximately 10 cm on a side and of good optical quality—can often be had as surplus materials for a very minimal price.

KDP is relatively stable and can be heated and cooled. ADP, on the other hand, decomposes when it is heated to temperatures of around 125°C, and when cooled it tends to crack.

Isomorphs of these materials have similarly been used in nonlinear optics, the most widely known isomorph being deuterated KDP, which is normally designated as KD*P. Some of the other isomorphs have been used because the temperature dependence of their refractive index allowed 90° phase matching for particular interactions.

The greatest attributes of ADP and KDP as device materials probably lie in two factors. The materials show considerable resistance to laser damage both of the mechanical type that occurs with very high power densities and the refractive-index change damage that occurs from low continuous-wave powers, although both effects are observed. In addition, the materials are easy to grow from aqueous solution which makes large, high quality, single crystals readily available.

Opposing these advantages, there are several disadvantages. The materials have poor infrared transmission because of the hydrogen atoms in their structure, and their refractive indices are fairly low, typically 1.50 to 1.55, so that they also have small nonlinear coefficients. Point group $\bar{4}2m$ allows three nonzero coefficients, d_{14}, d_{25}, and d_{36}. Of these, the first two are required to be equal by crystal symmetry, and the latter is expected to be equal to the other two by the Kleinman symmetry conjecture. Thus, for experimental purposes, only one coefficient has to be quoted.

The nonlinear coefficients of ADP and KDP (Appendix 2) have been measured with great care, since they have become the references against which most other materials are compared.

The refractive indices[172] at room temperature are given in Table 4.3. The

TABLE 4.3a   REFRACTIVE INDICES OF ADP AT ROOM TEMPERATURE.*

Wave-length (μ)	Index in air		Absolute index	
	Ordinary ray	Extraordinary ray	Ordinary ray	Extraordinary ray
0.2000	1.621996	1.563315	1.622630	1.563913
0.3000	1.545084	1.497691	1.545570	1.498153
0.4000	1.524035	1.479814	1.524481	1.480244
0.5000	1.514498	1.472068	1.514928	1.472486
0.6000	1.508851	1.467856	1.509274	1.468267
0.7000	1.504817	1.465193	1.505235	1.465601
0.8000	1.501508	1.463303	1.501924	1.463708
0.9000	1.498514	1.461830	1.498930	1.462234
1.0000	1.495628	1.460590	1.496044	1.460993
1.1000	1.492730	1.459481	1.493147	1.459884
1.2000	1.489751	1.458443	1.490169	1.458845
1.3000	1.486645	1.457436	1.487064	1.457838
1.4000	1.483381	1.456437	1.483803	1.456838
1.5000	1.479938	1.455427	1.480363	1.455829
1.6000	1.476302	1.454395	1.476729	1.454797
1.7000	1.472459	1.453333	1.472890	1.453735
1.8000	1.468400	1.452234	1.468834	1.452636
1.9000	1.464118	1.451093	1.464555	1.451495
2.0000	1.459603	1.449906	1.460044	1.450308

* After Zernike, *J. Opt. Soc. Am.*, **54**, 1215 (1964).

TABLE 4.3*b* REFRACTIVE INDICES OF KDP AT ROOM TEMPERATURE.*

Wave-length (μ)	Index in air		Absolute index	
	Ordinary ray	Extraordinary ray	Ordinary ray	Extraordinary ray
0.2000	1.648418	1.587119	1.649073	1.587740
0.3000	1.563459	1.512300	1.563953	1.512769
0.4000	1.540328	1.492165	1.540782	1.492601
0.5000	1.529833	1.483369	1.530271	1.483792
0.6000	1.523589	1.478476	1.524018	1.478892
0.7000	1.519097	1.475266	1.519522	1.475679
0.8000	1.515384	1.472875	1.515808	1.473285
0.9000	1.512006	1.470906	1.512428	1.471315
1.0000	1.508730	1.469155	1.509153	1.469563
1.1000	1.505428	1.467509	1.505851	1.467917
1.2000	1.502023	1.465899	1.502447	1.466307
1.3000	1.498465	1.464284	1.498891	1.464691
1.4000	1.494721	1.462634	1.495148	1.463042
1.5000	1.490766	1.460932	1.491195	1.461339
1.6000	1.486584	1.459162	1.487015	1.459570
1.7000	1.482161	1.457316	1.482594	1.457725
1.8000	1.477485	1.455386	1.477920	1.455795
1.9000	1.472547	1.453365	1.472985	1.453775
2.0000	1.467339	1.451249	1.467780	1.451660

* After Zernike, *J. Opt. Soc. Am.*, **54**, 1215 (1964).

indices can be computed from a Sellmeier equation of the form $n^2 = A + B^2/(1 - v^2/C) + D/(E - v^2)$, where the values of the constants A, B, C, and D are as given in Table 4.4. The refractive indices are temperature sensitive, and this property has been used to achieve x-y plane phase matching for the 5145 Å line of the argon ion laser in KDP and ADP, for frequency doubling to 2572 Å. The details of the temperature dependence of the indices have been studied independently by Adams and Barrett[1] and by Phillips.[136] The latter's conclusions are presented in the form of an equation for the change in

TABLE 4.4 SELLMEIER CONSTANTS FOR ADP AND KDP.*

A	2.133831	2.260476	2.164692	2.304082
B	8.653247×10^{-11}	1.011279×10^{-10}	9.633312×10^{-11}	1.114773×10^{-10}
C	8.134538×10^{9}	7.726552×10^{9}	7.691000×10^{9}	7.542305×10^{9}
D	8.069838×10^{5}	3.249268×10^{6}	1.479865×10^{6}	3.774363×10^{6}
E	2.500000×10^{5}	2.500000×10^{5}	2.500000×10^{5}	2.500000×10^{5}

* After Zernike, *J. Opt. Soc. Am.*, **54**, 1215 (1964).

index Δn from room temperature to the new temperature T so that

$$n(T) = n(298°K) + \Delta n \cdot \Delta T$$

$$\Delta T = (298 - T)°K$$

$$\Delta n = (n^2 + an + b)c \cdot \Delta T \tag{4.9}$$

The values of the constants, a, b, and c are given in Table 4.5 for the ordinary and extraordinary indices for both ADP and KDP and also for deuterated

TABLE 4.5 TEMPERATURE DEPENDENCE OF THE REFRACTIVE INDICES OF ADP, KDP, AND KD*P.†

	a	b	$c°K^{-1}$
ADP			
ordinary	−3.0297	2.3004	0.713×10^{-2}
extraordinary	0	0	0.675×10^{-6}
KDP			
ordinary	0	−1.432	0.402×10^{-4}
extraordinary	0	−1.105	0.221×10^{-4}
deuterated KDP			
ordinary	0	−1.047	0.228×10^{-4}
extraordinary	0	0	0.955×10^{-5}

† After Phillips, *J. Opt. Soc. Am.*, **56**, 629 (1966).

KDP. The constants are not significantly wavelength dependent, and in all cases the refractive indices increase as the crystals are cooled.

4.8 LITHIUM IODATE

Lithium iodate ($LiIO_3$) was first described in 1969 by two German workers, Nath and Haussuhl,[124] following an interest in the iodates that stemmed from earlier studies of alpha-iodic acid (HIO_3).[96] Lithium iodate offers the great advantages over alpha-iodic acid of mechanical stability over a wide temperature range (20–256°C) and relative freedom from degradation in a normal room environment. The point group is 6 and the crystal is negative uniaxial and phase matchable. The published refractive-index data are summarized in Table 4.6. The nonlinear coefficients have been measured by several workers, and the figures that are generally agreed upon, after some early dispute, are given in Appendix 2.

TABLE 4.6 REFRACTIVE
INDICES OF LITHIUM IODATE
AT ROOM TEMPERATURE.*

A	n_0	n_e
4000	1.948	1.780
4360	1.931	1.766
5000	1.908	1.754
5300	1.901	1.750
5780	1.888	1.742
6900	1.875	1.731
8000	1.868	1.724
10600	1.860	1.719

* After Nath and Haussuhl, *Appl. Phys. Lett.*, **12**, 186 (1968).

The device interest in lithium iodate exists because the compound's non-linear coefficients are closely comparable to those of lithium niobate, although it does not suffer from the refractive-index damage problems that plague the niobate. Consequently, despite other properties that are in some ways less desirable, lithium iodate has found application in efficient intra-cavity SHG from the Nd:YAG laser. Since the refractive indices are very stable with respect to temperature,[124] phase matching is achieved by angle tuning. The phase-matching angle for 1.084-μ radiation from a helium–neon laser is 28.9° from the c axis; for 1.1523-μ radiation it is 27.2°. It is interesting to note that because of its damage resistance, lithium iodate is used to generate the frequency-doubled radiation from a Nd:YAG laser used to pump the first commercially available parametric oscillator.

4.9 PROUSTITE

Crystals of proustite were first synthesized in large optical quality samples by the Royal Radar Establishment in England.[11,79,80] Proustite was developed after studies of the naturally occurring crystals found in mineral deposits showed that it is birefringent and acentric and has a good transmission band, stretching from about 6000 Å to beyond 13 μ. The material has large birefringence and, as such, is the only good quality synthetic crystal available that allows the study of phase-matched interactions between the region 8 and 13 μ (an atmospheric window of considerable interest) and the visible range.

Like lithium niobate, the crystal belongs to the point group 3m. However, its properties are quite different. The chemical formula of proustite is Ag_3AsS_3,

TABLE 4.7   REFRACTIVE INDICES OF PROUSTITE
AT 20°C*

(μm)	n_E	n_0
0.5876	2.7896	—
0.6328	2.7391	3.0190
0.6678	2.7094	2.9804
1.014	2.5901	2.8264
1.129	2.5756	2.8067
1.367	2.5570	2.7833
1.530	2.5485	2.7728
1.709	2.5423	2.7654
2.50	2.5282	2.7478
3.56	2.5213	2.7379
4.62	2.5178	2.7318

* After Hulme et al., *Appl. Phys. Lett.*, **10**, 133 (1967).

and formation occurs by the compounding of silver and arsenic sulfides. The crystal is negative uniaxial with a birefringence of $n_0 - n_e = 0.2$. The published refractive-index data are summarized in Table 4.7. Sellmeier equations of the forms[79]

$$n_0^2 = 9.220 + \frac{0.4454}{\lambda^2 - 0.1264} - \frac{1733}{1000 - \lambda^2}$$

and

$$n_e^2 = 7.007 + \frac{0.3230}{\lambda^2 - 0.1192} - \frac{660}{100 - \lambda^2}$$

can be used to calculate the indices between 0.6 and 20 μ at 20°C. Here λ is in microns.

From Miller's rule, we would expect that the high refractive indices of proustite would indicate a large nonlinearity, and this is so. The measured values are $d_{22} = 50 \, d_{36}$ (KDP) and $d_{31} = 30 \, d_{36}$ (KDP). The coefficient d_{33} was not measured (it is of little experimental interest), and the Kleinman condition was assumed ($d_{15} = d_{31}$).

Because of the exceptionally wide transmission band offered by proustite, the crystal has been used in several experiments involving the mixing of 10.6-μ radiation from a carbon dioxide–nitrogen laser with a visible laser, typically ruby. The material has been grown in single crystals of apparently perfect quality with dimensions of several centimeters. The material looks and handles very much like arsenic trisulfide glass. The mechanical properties of proustite seem to be similar, as well, and it has proved to be difficult to coat it with dielectric layers or to polish it to a high quality finish. Proustite has also exhibited instabilities to high-power laser beams, varying from degradation of the surface finish and minor internal damage to ignition!

Pyrargyrite (Ag_3SbS_3) is a material closely related to proustite that has also been studied to a limited extent.[79,80] Its linear and nonlinear optical properties are similar to those of proustite, although the absorption edge for this material occurs further into the red region, at approximately 7000 Å.

Cinnabar (HgS) is another material that has received some detailed study; it is closely related to proustite in its device properties, but not in its crystal or chemical properties.[30] This crystal belongs to point group 32; it transmits over essentially the same wavelength range as proustite and has similar sized nonlinear coefficients and refractive indices. It is, however, positive uniaxial.

The refractive indices of proustite, pyrargyrite, and cinnabar are essentially constant with respect to temperature, and therefore angle phase matching is used. Furthermore, because of the large birefringence, phase matching usually occurs at a small angle to the c axis, and hence walk-off between the ordinary and extraordinary beams can be a serious problem. For example, in an experiment on the conversion of $10.6\text{-}\mu$ radiation to the near-visible range by mixing it with the 6328-Å radiation from a helium–neon laser to generate the difference frequency, Boyd[30] quoted the maximum length of crystal for efficient interaction without walk-off limitation as being of the order of 0.06 cm of HgS, in an interaction in which all the beams are in the TEM_{00} mode. However, it is also worth noting that for some potential applications, this walk off effect may be a less serious limitation. We discuss such cases in Chapter 6.

5

Second-Harmonic Generation

5.1 INTRODUCTION

Second-harmonic generation (SHG) has been of practical interest ever since it was demonstrated that efficient conversion from fundamental to second-harmonic frequencies could be obtained at reasonable fundamental power levels.[70,43,48,63,125] This possibility has made available powerful sources of coherent radiation at hitherto unattainable wavelengths. (Table 5.1 lists some efficient frequency-doubled sources and their wavelengths.)

The most extensively studied conversion process of all has been the doubling of the 1.06 μ line obtained from the neodymium ion in various hosts. In particular, doubling of the continuous-wave Nd:YAG laser[63] has recently been the subject of intensive study, because the laser itself is both powerful and efficient, and because the green light obtained by doubling is well placed spectrally for detection by photomultipliers.

We begin this chapter by examining the relations obtained previously and putting in numbers to obtain a feel for the performances that are obtainable without resort to special techniques for increasing the efficiency of SHG. Subsequently we investigate some special techniques in greater detail.

5.2 PLANE WAVE INTERACTIONS

Chapter 2 contained a detailed discussion of phase-matched second-harmonic generation from a single-frequency plane wave. For the small-signal approximation we derived equation 2.52 for the second-harmonic power generated. The exact mathematical solution to the rate equations (equations 2.51) for the phase-matched case was given in equations 2.68 to 2.70, from which an expression for the generated power can easily be derived.

To illustrate these results we substitute numerical values for a case of practical interest—the conversion of the neodymium laser wavelength, 1.06 μ, to its second harmonic at 5300 Å in a crystal of lithium niobate. We assume

102

TABLE 5.1 SOME EFFICIENT SOURCES OF SECOND HARMONIC

Fundamental Wavelength Microns	Fundamental Power Watts	Second Harmonic Power Watts	Conversion Efficiency	Non-Linear Crystal	Reference
1.06 Nd Glass	—	10^4 pulsed	70% Peak 51% Average	KDP	70
1.06 Nd YAG	0.6 to 0.1 in Q-switched pulses. 0 - 11 KHz	0.9 to 0.3	Strong function of repetition rate	$LiIO_3$ Intra-cavity	43
1.06 Nd YAG	1.1 c.w.	1.1 watts c.w.	100%	$Ba_2NaNb_5O_{15}$ Intra-cavity	63
0.5145 Argon Ion	0.83	0.415	50%	KDP and ADP Intra-cavity	48
0.6943 Ruby	130.10^6 per cm^2	—	40%	$LiIO_3$	125

a crystal 1 cm long and an interaction phase matched in the x-y plane. (This occurs for phase-matching temperatures in the range 0–200°C, the exact value depending on crystal composition.) We further assume an area of beam cross section of 1 cm². The result appears in Figure 5.1, where we

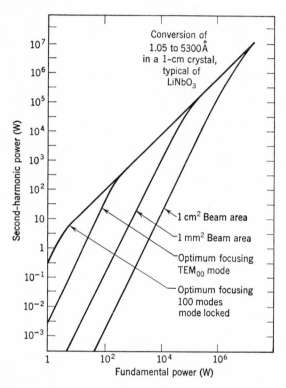

Figure 5.1 Theoretical curves for second-harmonic power versus fundamental power for various conditions. The crystal is taken to have parameters typical of lithium niobate 1 cm long. The laser source is taken to be at about 1 μ.

note that to obtain efficient conversions (10% or more), fundamental powers of 4×10^5 W or more are required. Figure 5.2 shows how the power at the second harmonic saturates as the fundamental begins to be depleted. Ideally 99% of the power is transferred from the fundamental to the second harmonic when the length in the crystal L is equal to about 3 times the characteristic length l_{SH}. To give some idea of what this corresponds to, suppose that $P[\omega] = 10^6$ W/cm². In this case, in the same example, $l_{SH} \approx 3$ cm.

The example just given represents an idealized situation, insofar as we assumed that:

1. The laser fundamental was single frequency, such that it could be described by an electric field of the form $\mathscr{E}(\omega) \cos (\omega t + \mathbf{k} \cdot \mathbf{R})$.

2. The wave was of infinite extent perpendicular to the propagation direction $\mathbf{k}$.

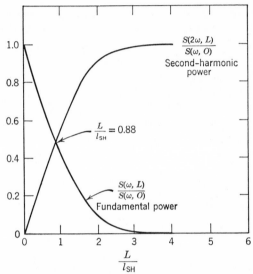

Figure 5.2 Form of the theoretical curve for the second-harmonic power from a long crystal, showing the effect of fundamental depletion.

5.3 FINITE BEAM SIZE

The analysis so far has considered fundamental beams of infinite extent, but the assumption is clearly not very realistic. However, it is a trivial matter to alter equation 2.52 to the case of a beam of area A cm² and power $W(\omega)$ to obtain for the second-harmonic power $W(2\omega)$ (in c.g.s. units)

$$W(2\omega) = \frac{512\pi^5 d^2 L^2 W^2(\omega)}{n(2\omega)n^2(\omega)\lambda^2 c A}$$

Figure 5.1 shows the result of making $A = 0.01$ cm² which, compared with $A = 1$ cm², naturally yields a gain in $W(2\omega)/W^2(\omega)$ of 100. This, however, assumes that the area A is sufficiently large that, throughout the length of the mixing crystal, the laser beam has approximately the same diameter. Clearly if A is made too small, diffraction will cause the beam to diverge, and the foregoing equation will no longer be valid. We must also remember that for any phase matching other than in the crystalline x-y plane, the fundamental and second-harmonic beams physically separate from each other, since they are necessarily of opposite polarizations. For a uniaxial crystal, the amount of this walk-off is readily assessed from the condition relating the wave normal and ray directions (equation 1.29). Thus, as long as we make sure that neither the walk-off nor the divergence is dominant, reducing the beam area or focusing the laser beam into the crystal substantially increases the efficiency of SHG.

Hence we see the need for an optimum degree of focusing. If the laser is focused too lightly, the beams will not be sufficiently concentrated for efficient operation. If the beam is too tightly focused, efficiency may be limited by the excessive beam divergence so produced, as well as by the physical separation of the fundamental and second-harmonic beams.

The detailed analysis of this situation is a complex matter, and an exact general solution has not been given for the case of propagation in a uniaxial crystal. However, in an exhaustive examination of the problem, Boyd and Kleinman[29] showed that it can be simplified in several ways, so that numerical results can be obtained for any case of interest by using graphical data from computer solutions. In particular, they demonstrated that for phase matching in a uniaxial crystal at some finite angle to the x-y plane, there are optimum values for crystal length and focal spot radius, once the crystal's parameters are known and the wave lengths are specified.

The same authors also showed that there is no optimum length for a crystal in which phase matching is possible in the x-y plane (e.g., lithium niobate) although for a given length there is an optimum focal-spot radius. In this situation they expressed the second-harmonic power in c.g.s. units in the following form:

$$W(2\omega) = \frac{1024\pi^5 d^2 L[1.068 W^2(\omega)]}{n(\omega)n(2\omega)\lambda^3 c} \tag{5.1}$$

The optimum spot radius for the Gaussian beam is w_0, where the beam profile is described by the function $\exp(-r^2/w_0^2)$ and the quantity w_0 is related to the crystal length through the following relations:

$$w_0 = \sqrt{b\lambda/2\pi}$$
$$b = \frac{L}{2.84} \tag{5.2}$$

The terminology used here is commonly employed in the description of single transverse mode laser beams, where b is the confocal parameter* for a gas laser producing a lowest order mode described by w_0. At a distance

* In a confocal resonator b = radius of curvature of the mirrors = spacing of the mirrors.

z away from the focus, such a beam has a spot radius w_0^1, where

$$w_0^1 = w_0\sqrt{1 + 4z^2/b^2} \tag{5.3}$$

Either of the parameters w_0 or b, together with the wavelength and the position of the focus, is sufficient to specify the beam profile or the intensity at any position in space. These relations were first derived by Boyd and Kogelnik[25] in the consideration of the modes of propagation for light in the typical laser resonator. The techniques for matching the mode from a laser of given w_0 and b into another mode having different w_0' and b' by the use of suitably chosen and placed lenses have been treated in detail by Collins.[44]

If we now take equation 5.2 and again substitute numerical values representative of a 1-cm crystal of lithium-niobate, with a fundamental wavelength of 1.06 μ, we obtain the second result shown in Figure 5.1, illustrating the dramatic increase in conversion efficiency that can be obtained by the use of a suitably focused beam. This technique has been used, in conjunction with others, to obtain very high mean conversion efficiencies from low mean power continuous-wave lasers. We discuss this later, along with practical results and applications.

The foregoing result applies strictly to single transverse mode laser beams. In the more general case of a multimode laser, two approaches are possible. If the mode structure is known in a mathematically expressible form, a calculation similar to that of Boyd and Kleinman[29] could be performed for the case of interest. More often, however, only the gross parameters of the laser beam such as its beam divergence and area at some point in space are known, and the exact phase distribution across the beam area is unknown. Under these conditions it is useful to estimate the tolerance of phase-matched SHG to divergent beams and possibly to sources of great spectral linewidth. The phase-matching condition $\Delta k = k_{2\omega} - 2k_\omega = 0$ is satisfied only for a single wavelength for each direction in the crystal. However, the SHG varies not as $\delta(\Delta k)$ but as $\sin^2 (\Delta K L/2)/(\Delta k L/2)^2$, and this allows a small tolerance $\Delta k = \pm \pi/L$, where L is the length of the crystal. This in turn relates to a small angular tolerance, which can be calculated using equations 3.11 or 3.12. In like manner we can estimate the effect of a small change in frequency of the source by calculating $\partial(\Delta k)/\partial\omega = \pi/(L \cdot \Delta\omega)$

$$\Delta\omega = (\pm) \frac{\pi}{L}\left(\frac{\partial k(2\omega)}{\partial\omega} - \frac{\partial k(\omega)}{\partial\omega}\right)^{-1}$$

For a broad-band fundamental source, the power at the "second harmonic" does not consist only of each of the fundamental frequencies doubled; it also contains combinations [i.e., if the source contains ω and $\omega + \Delta\omega$, then at the second harmonic we obtain 2ω, $2\omega + \Delta\omega$, and $2(\omega + \Delta\omega)$]. This circumstance leads to additional effects, which we examine next.

5.4 EFFECTS OF MODE STRUCTURE ON SHG

A very small proportion of lasers in use today operate in true single mode, that is to say, with a single frequency and a wavefront with a Gaussian intensity profile. Most gas lasers have the Gaussian intensity profile of the TEM_{00} mode, but examination of their frequency spectrum indicates that they produce a series of discrete frequencies spaced by $c/2L$ Hz, where c is the velocity of light and L is the length of the laser resonator. In general, for each of these "longitudinal modes" there is often a multitude of transverse modes, and the intensity profile of the output is not Gaussian, nor does it have a simple phase distribution across it. Such complex frequency and phase distributions of the available power have a surprising effect upon SHG, since certain conditions increase the efficiency with which a given amount of power is converted to second harmonic. At the same time, the random and time-varying phase relations between modes cause the second-harmonic power to fluctuate in a noiselike manner, even though the fundamental power remains steady.

We can see intuitively why this mode structure leads to an increase in a mean generation rate if we bear in mind that the second-harmonic electric field depends upon $E(\omega)^2$ rather than $E(\omega)$. The time output of the multimode laser looks noiselike, but it has a repetitive structure that repeats every $2L/c$ sec and only changes its detailed form over many cycles as the relative phases of the modes drift (see Figure 5.3). The SHG process naturally gains more from the peaks of this noiselike wave form than it loses from the troughs, and on the average a gain in conversion is obtained. A broad analysis of this effect was first presented by Ducuing and Bloembergen.[50] We take a simpler theory, which illustrates the mechanism of the enhancement but simplifies the notation.

Assume that the laser output can be described as a sum of modes, uniformly spaced in frequency and of essentially equal amplitude, and having the same intensity and phase profiles. Such a description fits most gas lasers and many solid-state lasers operating in the TEM_{00} mode. Thus

$$E_{\mathrm{F}} = \sum_{n=1}^{N} \mathscr{E}_n e^{i(\omega_n t - \varphi_n)}$$

$$= \sum_{n=1}^{N} E_n e^{i\omega_n t}$$

where $E_n = \mathscr{E}_n e^{-i(\varphi_n)}$ and the integer n runs from 1 to N. The laser power is then given by $S(\omega) \approx \sum_1^N E_n E_n^* = N \cdot E_n \cdot E_n^*$. The second-harmonic

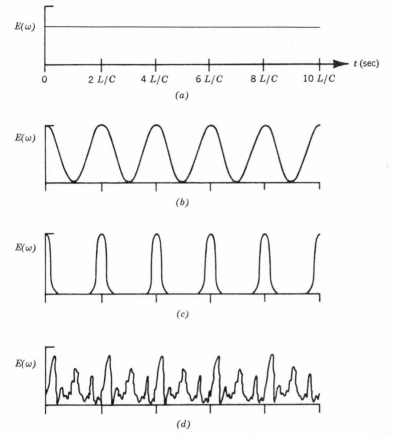

Figure 5.3 The effects of multiple modes on the temporal output of a laser: (*a*) Single mode stabilized, (*b*) two stabilized modes, (*c*) multiple modes phase locked, (*d*) multiple modes randomly phased.

electric field is given by an expression of the form

$$E(2\omega) = K \sum_{n=1}^{N} \sum_{m=1}^{N} E_n E_m e^{i(\omega_n + \omega_m)t}$$

and the intensity due to the multimode field, as would be seen by a square-law detector such as a photomultiplier, is of the form

$$I(2\omega)_{mm} = K^2 \sum_{n=1}^{N} \sum_{m=1}^{N} \sum_{o=1}^{N} \sum_{p=1}^{N} E_n E_m E_o^* E_p^* e^{i(\omega_n + \omega_m - \omega_o - \omega_p)t}$$

We assume perfect phase matching for all combinations of fundamental modes.

Of the multitude of frequency components contained within the expression for $I(2\omega)_{mm}$, only those which fall within the bandwidth of the detector are registered. Thus we set the condition that the signal to be detected, $I(2\omega)$ is that obtained under the condition $\omega_n + \omega_m - \omega_o - \omega_p = 0$. We can also select, from the general sum, terms which are phase dependent and those which are not. The phase of the term $E_n E_m E_o^* E_p^*$ is $\varphi_n + \varphi_m - \varphi_o - \varphi_p$, and thus the sum is phase independent only when $n = m = o = p$, when $n = o$ and $m = p$, and when $n = p$ and $m = o$. Thus we have for the phase-independent term

$$\langle I(2\omega)_{pi} \rangle = K^2 \sum_1^N (E_n E_n^*)^2 + 2K^2 \sum_{n=1}^N \sum_{m=1(n \neq m)}^N E_n E_n^* E_m E_m^*$$
$$= K^2 [N + 2N(N - 1)](E_n E_n^*)^2$$

By specifying that the summation include only phase-independent terms, we select the mean time-averaged intensity. The remaining terms in the summation that satisfy the condition on the frequencies but that include noncanceling random phases are slowly time varying in their relative amplitudes because of the random phase interference effects. The frequency spectrum of these terms is composed of frequencies given by $\Delta\omega_n = (\partial\varphi_n/\partial t)$. This contrasts with the next case we consider (Section 5.5), in which the phases are not randomly varying and can, therefore, be allowed in the summation.

Now, had we used the power at ω in a single frequency to generate second harmonic, we would have had

$$\langle I(2\omega) \rangle_{sm} = K^2 N^2 (E_n E_n^*)^2$$

from which

$$\frac{\langle I(2\omega)_{pi} \rangle}{\langle I(2\omega) \rangle_{sm}} = 2 - \frac{1}{N}$$

For a large number of modes, this solution corresponds to an enhancement of 2 in the mean power generated. However, there remain the terms in the second-harmonic power which depend critically on the relative phase between modes, and these give rise to noiselike fluctuations on top of the mean power listed earlier. Ducuing and Bloembergen estimated the value of the mean rms fluctuation in second-harmonic power at 18% for $N = 3$, and 25% for $N = 4$; for large N, they found that this value is given by $\sqrt{\frac{2}{3}N}$. Thus we realize that although use of a multimode laser can boost the mean conversion efficiency, it also introduces into the output a randomness that can be troublesome in many applications.

Since, in practice, we are usually looking for all the efficiency possible in SHG, we are often forced to seek other ways of boosting the SHG efficiency; one such way is found in the use of a mode-locked laser source.

5.5 SHG FROM A MODE-LOCKED LASER

The unique feature of a mode-locked laser is that at regular points in time, spaced by $2L/c$ sec, all the longitudinal modes come into phase with one another. Thus we can describe the electric field of such a laser in the following form:

$$E_F = \sum_{n=1}^{N} E_n e^{i(\omega + n\Delta\omega)t}$$

$$\Delta\omega = \frac{\pi c}{L}$$

where the E_n are all real and (we assume for our purpose here) of similar amplitude. Thus at $t = 0$, all are in phase and, likewise, every $t = 2L/c$. This makes the laser output appear as a train of uniformly spaced pulses with a ratio of spacing between pulses to pulsewidth of about N and with a spacing $2L/c$ sec. Since SHG depends on the mean of the square amplitude rather than the square of the mean, we can expect this pulselike form to have a marked effect on the efficiency of conversion. Our analysis proceeds as follows.

Given that the power output of the laser is $S'(\omega) \approx \langle E_f^2 \rangle$, we have the relation as before, namely, $\langle E_f^2 \rangle = N \cdot \langle E_n^2 \rangle$. The second-harmonic field strength, assuming perfect phase matching for all combinations of modes, is given by

$$E(2\omega)_{ml} = K \sum_{n=1}^{N} \sum_{m=1}^{N} E_n E_m e^{i[2\omega + (n+m)\Delta\omega]t}$$

and the second-harmonic intensity by the relation

$$I(2\omega)_{ml} = K^2 \sum_{n=1}^{N} \sum_{m=1}^{N} \sum_{o=1}^{N} \sum_{p=1}^{N} E_n E_m E_o^* E_p^* e^{i(n+m-o-p)\Delta\omega t}$$

Since we are interested in the average value of this intensity, we must perform the summation and calculate the average value of $I(2\omega)$ which is given by

$$\langle I(2\omega) \rangle_{ml} = K^2 E_n^4 \sum_{n=1}^{N} \sum_{m=1}^{N} \sum_{o=1}^{N} \sum_{p=1}^{N} \delta(n + m - o - p)$$

where δ is the Kronecker symbol (see Section 2.9). This summation can be split into two sums of the form

$$\sum_{n=1}^{N} \sum_{m=1}^{N} \sum_{o=1}^{N} \sum_{p=1}^{N} \delta(n + m - o - p)$$

$$= \sum_{J=1}^{N} J^2 + \sum_{J=N+2}^{2N} (2N - J + 1)^2$$

$$= \sum_{J=1}^{N} J^2 + \sum_{J=1}^{N-1} J^2$$

$$= \frac{N(N + 1)(2N + 1)}{6} + \frac{(N - 1)N(2N - 1)}{6}$$

Thus we obtain for the average second-harmonic power from the mode-locked laser the following result:

$$I(2\omega)_{ml} = K^2 E_F^4 \frac{N(2N^2 + 1)}{3}$$

where

$$\frac{\langle I(2\omega)_{ml} \rangle}{\langle I(2\omega)_{sm} \rangle} = \frac{2N^2 + 1}{3N}$$

For a large number of modes N, this clearly reduces to $\frac{2}{3}N$. For a neodymium glass laser, N may be as large as 1000. It should be remembered, however, that we have assumed that phase matching is maintained for the whole of the fundamental frequency bandwidth. In practice, this often proves to be a serious limitation.[45]

For the continuous-wave Nd:YAG laser, N is typically of the order of 10^2. Thus for a fully mode-locked TEM_{00}-mode, 1-W laser with a lithium-niobate crystal 1 cm long, and having optimum focusing, we can expect to generate on a continuous basis approximately 90 mW of green light. This represents an efficiency of only about 10%, despite a factor of 66 gain through mode locking.

5.6 INTRACAVITY SHG

We have seen in the foregoing sections that it is relatively simple to generate small amounts of second-harmonic power on a continuous-wave (CW) basis, and that in pulsed operation, high conversion efficiencies can be obtained readily. However, since many applications of SHG require CW or quasi-CW power, we are forced to search for ways to increase the CW conversion efficiency. Brief reflection suggests that it might be better to place the SHG crystal within the laser resonator, where it would be subjected to the high circulating power contained within the high-Q resonator. Furthermore, since the optimum loss rate or output coupling for most CW lasers is of the order of 1 to 4%, we might expect that, if we can introduce a SHG loss rate or conversion of this order in place of an output mirror transmission of 1%, we might again achieve an optimum coupling situation and with luck (or good design) obtain in second-harmonic power effectively all the power that would otherwise have been available as fundamental output. This case has been examined both theoretically and experimentally by several workers.[137,149] We now outline a theoretical analysis that lucidly illustrates the features of this type of SHG.

In order to show the presence of an optimum coupling situation for intracavity SHG, it is necessary to include in the theoretical description certain detailed parameters concerning the laser. Indeed, the whole operation of this type of SHG depends critically on striking the right balance between the various loss and pump rates associated with the laser optical field within the resonator. Figure 5.4 illustrates the various critical factors. The laser rod or

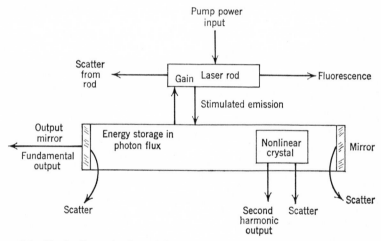

Figure 5.4 Block schematization of the energy flow and storage mechanisms that must be considered in an analysis of intracavity SHG.

discharge tube provides a source of power to drive the oscillator. The power is coupled into the resonator or optical energy reservoir through the medium of the photon flux. The photon flux can then leak out of the resonator by various routes, such as, through an end mirror to provide the normal laser output, by scatter or mismatch reflection from any of the elements, by linear absorption, or by conversion to second harmonic in a nonlinear element. In the normal laser situation, the laser flux starts to build up as soon as the gain of the medium exceeds the fixed losses. Since the loss rate is usually proportional to the photon flux, the power loss continues to increase as the flux grows, until the loss exactly equals the power input rate. The power input rate itself is finite and can be expressed either in terms of a finite input rate to the inverted population of the laser or in terms of a flux-dependent gain–saturation parameter. We use the former approach, following closely the work of Polloni and Svelto.[137]

Two basic laser configurations concern us—four-level and three-level varieties, as illustrated in Figure 5.5. The outstanding difference between

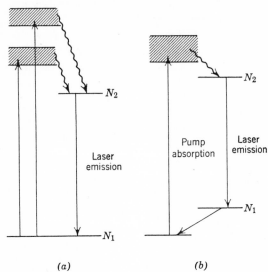

(a)   *(b)*

Figure 5.5 Schematic energy level diagrams for: (a) the three-level laser, (b) the four-level laser.

them is that in the three-level laser, the laser transition terminates in the ground state, whereas in the four-level configuration, it terminates at a level above the ground state. If there are N_2 ions in a given volume of laser material excited into the upper laser transition level, and there are N_1 ions in the lower level, the gain is proportional to $N_2 - N_1$ and the absorption at the laser frequency to $N_1 - N_2$. Thus there is a dramatic operational difference between the two systems. In the three-level laser, before any laser gain is produced at all, more than half of the ions must be excited. However, in the four-level laser (provided the lower laser level is well separated from the ground state so that it has a negligible thermal population), any excited ions N_2 lead to gain, since N_1 is always essentially zero.

Good examples of the two laser types are ruby (Cr^{3+} in Al_2O_3), which is the most common three-level laser, and Nd^{3+} in yttrium-aluminum-garnet, glass, or calcium tungstate, which are four-level lasers. Of the latter group, Nd:YAG is by far the most popular as a CW solid-state laser, and this material is almost exclusively used with continuous CW pumping, where the full advantage of the four-level system really shows.

A simple mathematical model can be used to describe the operation of the laser. It does not take into account multiple modes, mode competition, or spatial hole-burning effects, but it does give a good feel for the basic characteristics of the laser and of intracavity SHG. The model describes the laser action by two coupled-rate equations—one for the stored energy in the inverted population and the other for the stored energy in the photon flux. An energy input is provided for by a constant pump rate feeding the inverted population; the energy losses are accounted for by rate terms describing the loss of photons from the optical cavity, the loss of stored energy by fluorescence, and so on. The population inversion is given by a single parameter n, which is defined as $n = N_2 - N_1$. The total number of active ions in the laser rod is given by N, which for the three-level laser is equal to $N_2 + N_1$. If we now assume that the laser cavity is essentially filled with laser material, except where the nonlinear crystal is placed, we can write very simple rate equations to describe the laser action. For a four-level laser the rate equations are

$$\frac{\partial n}{\partial t} = \beta N - Bnq - \frac{n}{\tau} \tag{5.4}$$

$$\frac{\partial q}{\partial t} = Bnq - (K_i + K_o)q - K_{SH}q^2 \tag{5.5}$$

The three-level laser rate equations are

$$\frac{\partial n}{\partial t} = \beta(N - n) - 2Bqn - \frac{n + N}{\tau} \tag{5.6}$$

$$\frac{\partial q}{\partial t} = Bnq - (K_i + K_o)q - K_{SH}q^2 \tag{5.7}$$

In equations 5.4 to 5.7, β is a parameter proportional to pump rate and, generally, pump lamp power; N is the number of active ions in the laser material; B is the stimulated emission coefficient per photon and per excited ion; τ is the fluorescence decay time; K_i accounts for internal losses within the laser; K_o is the output loss through the laser mirrors; K_{SH} is the loss due to SHG, and q is the total number of photons in the laser resonant cavity.

From equations 5.4 and 5.6 we see that the difference between a three-level and a four-level laser affects only the rate equation for the population inversion. In the three-level laser rate equation, the emission of one quantum into the stored optical field is accompanied by a decrease in the upper level population N_2, of unity and an increase of N_1 by one. Consequently the inversion $N_2 - N_1$ changes by 2, which accounts for the factor of 2 appearing in the rate equation (5.6). The steady-state values (n_o, q_o) are obtained from equations 5.4 and 5.5 by setting the time derivatives to zero

$$\beta N - Bn_o q_o - \frac{n_o}{\tau} = 0$$

$$Bn_o - (K_i + K_o) - K_{SH} q_o = 0$$

(5.8)

To operate the laser as an efficient generator of harmonic power, we set $K_o = 0$ (i.e., we use high-reflectivity mirrors so that the only fundamental losses are due to scatter and other unavoidable losses). The second-harmonic power is then

$$W(2\omega) = K_{SH} q_o^2 \hbar\omega$$

(5.9)

From equations 5.8 and 5.7 we find

$$W(2\omega) = \frac{\hbar\omega}{4B^2\tau^2} \left\{ \sqrt{[(K_{SH} + B\tau K_i)^2/K_{SH}] + 4B\tau(\beta BN\tau - K_i)} - \frac{K_{SH} + B\tau K_i}{\sqrt{K_{SH}}} \right\}$$

(5.10)

The quantity $W(2\omega)$ has a maximum when $(K_{SH} + B\tau K_i)/K_{SH}$ is a minimum, that is, when

$$K_{SH} = B\tau K_i$$

(5.11)

This is an interesting result, since it shows that once the interaction is properly adjusted, the amount of second-harmonic power generated is the maximum that can be coupled out, independent of the pump rate. In contrast to this, by setting $K_{SH} = 0$ in equation 5.5 and going through the same procedure, it is easy to show that for the fundamental, the optimization of the output *does* depend on the pump rate, since

$$K_{o(opt)} = \sqrt{\tau\beta BN K_i} - K_i$$

(5.12)

By substituting equation 5.11 in equation 5.10 we find that the optimum second-harmonic power output is given by

$$W(2\omega)_{opt} = \frac{\hbar\omega}{B\tau} (\sqrt{\beta NB\tau} - \sqrt{K_i})^2$$

By using equation 5.12 in equations 5.4 and 5.5, the optimum amount of output at the fundamental can be shown to be equal to $W(2\omega)_{opt}$.

Thus in principle, the technique described makes it possible to obtain 100% of the useful output of the laser at the second harmonic. Note, however, that this does not mean that 100% conversion of fundamental to second harmonic has been obtained. In fact, the actual conversion within the laser resonator of the fundamental beam to the second harmonic is nearer to 1% than 100%. In other words, all the power that the laser could previously generate as output at the laser frequency is now generated, after modifications to the laser, as power at the harmonic frequency. The modifications have included reducing the output of the laser at the fundamental frequency to zero.

It is also interesting to note that the existence of an optimum value for the second-harmonic loss rate, K_{SH}, implies that as K_{SH} is increased, the second-harmonic power will go through a maximum, and too much coupling would be as bad as too little. Indeed, effects due to overcoupling have been reported by Geusic et al.[63] in an intracavity SHG experiment.

As a numerical example we consider the case of the Nd:YAG laser generating second harmonic in barium sodium niobate. For the Lorentzian laser line[137]

$$B = \frac{2c^3}{\omega^2 \Delta\omega\tau^{-2}A_L Ln_L{}^3} \tag{5.13}$$

where c = velocity of light (cm/sec)
ω = laser frequency (rad/sec)
$\Delta\omega$ = linewidth (rad/sec)
A_L = beam area in laser crystal = πw^2 (cm²)
L = length of laser cavity (cm)
n_L = refractive index of laser rod

The second-harmonic coupling coefficient is given by the relation

$$K_{SH} = \frac{64h\pi \, d^2\omega^3 L_C{}^2}{cn_C{}^3 n_L{}^2 w_0{}^2 L^2} \tag{5.14}$$

where d = nonlinear coefficient
h = Planck's constant
L_C = length of nonlinear crystal (cm)
n_C = refractive index of nonlinear crystal
w_0 = spot radius in the crystal

If we now combine equations 5.11 and 5.14 with 5.13 to obtain a value for the optimum length $L_{C(opt)}$ of the nonlinear crystal, we find

$$L_{C(opt)} = \sqrt{c^5 n_C{}^3 \alpha_i / 32\pi \, d^2 h\omega^5 \, \Delta\omega n_L{}^2} \, \frac{w_0}{w} \tag{5.15}$$

where α_i is the loss per pass in the resonator ($K_i = \alpha_i c/n_L L$) and w is the spot radius at the laser rod.

Substituting values typical of the Nd:YAG laser with $\alpha_i = 10^{-2}$, we obtain $L_{C(opt)}$ $(w/w_0) \approx 2$ cm. Thus, if we choose to design the optical resonator so that the mixing crystal is placed at a focus in the beam, we can arrange (w/w_0) to be about 4 or more, and we find values for optimum crystal length of 5 mm or less. A typical experimental layout appears in Figure 5.6, here

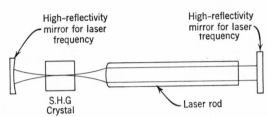

Figure 5.6   A typical experimental configuration for intracavity SHG showing the placing of the nonlinear crystal at the neck of the oscillating mode.

the laser mode comes to a focus in the SHG crystal but is large in the laser crystal.

Geusic et al.[63] have reported 100% conversion of the useful output of a Nd:YAG laser to the second harmonic by this technique. They used a barium sodium niobate crystal 3 mm long in a Nd:YAG laser nominally capable of giving 1-W output. It should be emphasized, however, that the second-harmonic is generated from a linearly polarized fundamental. Thus in a device like the Nd:YAG laser, which is normally unpolarized, only about 50% of the available power is converted to second harmonic even in the most favorable case. Geusic's comparison was made between the output of a Nd:YAG laser, polarized by a plane-parallel plate at Brewster's angle inside the cavity, and the intracavity SHB in the same laser.

In addition, it should be noted that equal amounts of the generated second-harmonic power travel forward and backward in the laser resonator. Thus 50% of the generated power is lost unless some measures are taken to collect it. Attempts to reflect the backward traveling power, so that it combines with the forward generated power to produce a single beam, lead to further complications because the two beams interfere with each other (these problems have been considered in great detail by R. G. Smith[149]). For unidirectional conversion from an unpolarized laser, therefore, the maximum conversion is about 25%.

In practice, severe problems are encountered in trying to achieve performance as good as 25%. Most obvious is our assumption that the act of inserting the nonlinear crystal into the laser resonator does not significantly increase the internal loss of the laser K_i. It is also implicit, although perhaps not obvious without a detailed calculation, that this total loss must be very low (of the order of 1 % per pass through the whole laser resonator). Thus exceptionally high quality crystals are needed. Also not stated, but implicit in the foregoing, is that the nonlinear crystal must not perturb the laser mode unduly. This requires that the material, in addition to having very low zero-flux insertion loss, shall have extremely low absorption at the laser and second-harmonic frequencies, since most of the nonlinear materials of prime interest for this type of application have very temperature-sensitive refractive indices. Thus a small absorption can lead to significant heating of the nonlinear material along the beam path. This disturbs the phase matching and also generates a lenslike distortion in the nonlinear crystal.

Nevertheless, despite the limitations on the availability of good doubling crystals, there are commercially available intracavity doubled YAG laser systems that yield relatively high efficiencies (10% or more). The system illustrated in Figure 5.7 shows a well-engineered continuous-wave Nd:YAG

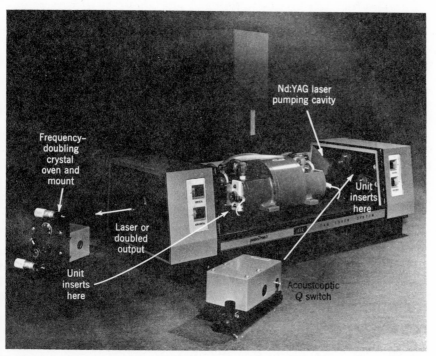

Figure 5.7 Laser system offering intracavity SHG. The laser operates at 1.06 μ (Nd:YAG) and is frequency doubled by the attachment shown, which is inserted into the laser cavity. It is a small, temperature-controlled oven containing a crystal of barium sodium niobate about 0.5 cm long. The acoustooptic Q switch allows Q-switching at a rate of up to about 10 KHz and may be used with the doubling accessory to obtain pulsed green output. Photo courtesy of Quantronix Corporation, 225 Engineers Row, Smithtown, N.Y.

laser with two accessories that are easily inserted into the laser cavity. The left-hand assembly is an oven containing a 5-mm cube of barium sodium niobate for frequency doubling the laser output to give radiation at 5300 Å. The right-hand assembly is an acoustooptic Q switch using a block of fused silica with an acoustic transducer attached, to Q-spoil the laser cavity at repetition rates of up to about 10 KHz, yielding output pulses of typically 10^{-6} sec length. The two attachments may be used individually or together, to yield pulsed or CW green or 1.06-μ radiation.

To return to the three-level laser, the rate equations (5.6 and 5.7) may be solved by a similar process to obtain expressions for optimum coupling for SHG or fundamental power output. The result so obtained for the optimum coupling for SHG when $K_0 = 0$ is given by the relation

$$K_{\mathrm{SH(opt)}} = \frac{2B\tau K_\mathrm{i}}{1 + \beta\tau}$$

Notice that here K_{SH} depends on the pump rate, unlike the equivalent four-level result. Thus the output coupling for optimum SHG is now pump-power dependent, and the optimum length for a doubling crystal is simply given by the relation

$$L_{\mathrm{C(opt)3\ level}} = \frac{2}{1 + \beta\tau}\ L_{\mathrm{C(opt)4\ level}}$$

Intracavity SHG with a three-level laser has been studied experimentally using a ruby laser and lithium iodate as the frequency-doubling material.[125] The niobates cannot be used in this type of experiment, since they all absorb at the second-harmonic frequency of the ruby laser (6943 Å doubled to 3471 Å).

5.7 PICOSECOND TIME DOMAIN MEASUREMENTS BY SHG

SHG has also found application in the study of optical pulses in the pico-second (10^{-12}) time domain.[6,104] This application exploits the nonlinearity inherent in the SHG process to perform auto- or cross-correlation on optical beams. In the simplest case, the picosecond pulse train to be studied is split into two equal components by a beam splitter (see Figure 5.8).

One of these components is passed through a variable time delay τ. The two beams are then recombined in a nonlinear crystal where second harmonic is generated in such a manner that the second-harmonic power is derived, not by doubling each component, but only by adding the direct and delayed components. Thus, if the input is described by $f(t)$, the delayed component is $f(t + \tau)$, and the second-harmonic component is $E(2\omega) = f(t) \cdot f(t + \tau)$.

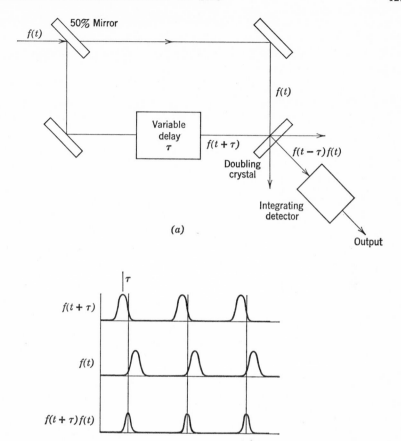

Figure 5.8 Picosecond pulse length measurement by SHG autocorrelation: (*a*) Typical experimental layout, (*b*) time sequence of the autocorrelation procedure.

If this is measured by a normal detector (photodiode or photomultiplier) whose bandwidth is much less than 10^{12} Hz, the photocurrent measured will be

$$I(2\omega) = \frac{\int [f(t) \cdot f(t + \tau)]^2 \, dt}{\int dt}$$

Thus the current is a function of τ and its dependence on τ describes the pulse shape. When τ is zero, the pulse height is measured. When τ is increased, the overlap between the direct and delayed beams decreases until, for large enough τ, there is no overlap and no second harmonic. Since a pulse 10^{-12} sec long is only 0.03 cm long, a τ large enough to produce no overlap can be readily produced by changing the path length.

Several techniques have been outlined for achieving the correct form of SHG in this application. The first, due to Giordmaine et al.,[104] appears in Figure 5.9. It uses noncollinear beams phase matched along the common

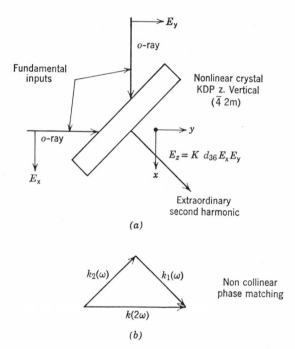

Figure 5.9   Use of noncollinear phase matching to achieve autocorrelation by bulk SHG.

k vector direction. The independent beams are then not phase matched. This technique suffers the disadvantage that interaction occurs over a large volume of nonlinear crystal, which limits the time resolution.

The alternative technique, due to Armstrong,[6] uses orthogonally polarized components interacting at the surface of a gallium arsenide wafer to give the second harmonic (Figure 5.10). In gallium arsenide the only nonlinear coefficient allowed by crystal symmetry is $d_{14} = d_{25} = d_{36} = d_{xyz}$. Thus if a (111)-oriented slice of gallium arsenide is used and the input beams are orthogonally polarized along the x and y directions respectively, the angle of incidence being 45°, then the SHG is polarized along the z direction, and thus

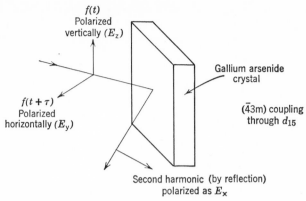

Figure 5.10   The use of orthogonal polarizations and a slice of gallium arsenide to achieve autocorrelation by surface SHG.

couples to an emerging wave in the plane of incidence whose polarization vector lies along the propagation direction of the incident beams.* Since for the case of greatest interest, that of the neodymium laser, the second harmonic is heavily absorbed by the gallium arsenide crystal, all the generation takes place within a very thin surface layer, and effects due to phase matching (or lack of it) are not significant. Furthermore, the very short interaction length makes possible time resolution of at least 4×10^{-13} sec.

The reader should be warned that this technique, in common with others, must be used with great care, since the repetitive nature of the pulse trains studied can lead to ambiguous results. The details of the conditions to be observed in interpreting the results of such measurements have been discussed in great detail in the literature.[162]

Finally, we note that the foregoing techniques can be extended to the general situation of cross-correlation by the use of two mode-locked beams, one to probe the other. If one input is $f(t)$ and the other $g(t)$, the photocurrent is given by

$$I_{(\text{sum})} = \frac{\displaystyle\int [f(t) \cdot g(t + \tau)]^2 \, dt}{\displaystyle\int dt}$$

The integration is performed by the limited time response of the detector. Then if the function $g(t)$ consists of a train of fast pulses slightly out of frequency register with the signal $f(t)$, sampling of the signal by the reference can occur, enabling the fast signal to be displayed through a much smaller bandwidth display system.

* This is an example of the SHG upon reflection mentioned in Section 2.19.

6

Parametric Up-Conversion

6.1 INTRODUCTION

SHG represents a special case of the more general process of sum-frequency generation in which two different frequencies are added to form the sum. The special feature of SHG is, of course, that the two sources are of the same frequency and, in fact, are normally the same beam, "mathematically" used twice. Consequently, it is normal to assume that both are of equal power or amplitude although, as we have seen, some interest can lie in situations in which the two source beams are experimentally separable (e.g., by polarization).

The more generalized sum-frequency generation has attracted little attention to date. The only commercial application seems to have been the generation of multiple new frequencies from combinations of laser, SHG, and parametric oscillator in attachments such as that offered by Chromatix* for their Laser/Doubler/Parametric Oscillator source. By forming the sum frequency of various combinations of lines from this unit, the system offers coherent, tunable radiation essentially covering the whole wavelength band from 2500 Å to beyond 3 μ.

The subject of this chapter, up-conversion, is another special case of sum-frequency generation. Its distinguishing feature is the boundary condition applied during the solution of the coupled equations 2.39. The frequency ω_3 is generated by the addition of two sources at ω_1 and ω_2. However, it is assumed that the power at the frequency ω_2 is much larger than the power at ω_1 and that initially the power at ω_3 is essentially zero: sum-frequency generation usually takes the powers at ω_1 and ω_2 as being comparable. Applying the former condition to the coupled equations 2.39, we see directly that we can set $\partial E_2/\partial z$ equal to zero and that, therefore, in the phase-matched

* Chromatix, 1145 Terra Bella Avenue, Mountain View, Calif. 94040.

case we can obtain a particularly simple solution, namely,

$$\mathscr{E}_3(z) = \left[\frac{\omega_3{}^2 k_1}{\omega_1{}^2 k_3}\right]^{1/2} \mathscr{E}_1(0) \sin \frac{z}{l_{\text{puc}}} \tag{6.1}$$

$$\mathscr{E}_1(z) = \mathscr{E}_1(0) \cos \frac{z}{l_{\text{puc}}} \tag{6.2}$$

Notice that the power at the frequency ω_1 is totally transferred to the sum-frequency beam ω_3 after a characteristic length $\pi l_{\text{puc}}/2$, where l_{puc} is given by the relation

$$l_{\text{puc}} = \left[\frac{4\pi d}{c^2}\left(\frac{\omega_1{}^2 \omega_3{}^2}{k_1 k_3}\right)^{1/2} \mathscr{E}_2\right]^{-1} \tag{6.3}$$

If we allow the interaction to continue (i.e., if the crystal is long enough), the power is transferred back and forth cyclically between ω_1 and ω_3 every distance of $\pi l_{\text{puc}}/2$. In principle, therefore, we have a means of frequency-shifting radiation with high efficiency, provided we can physically make the distance l_{puc} equal or comparable to the crystal lengths that are available. The particular interest in doing this, from a device point of view, stems from two factors.

First, there are now available very efficient low-noise detectors of visible radiation such as the human eye or photomultipliers, whereas most infrared detectors are very inefficient by comparison. For example, it is possible to detect a few quanta per second at 5000 Å (green), but at 10 μ the usual requirement for detection is about 10^8 quanta/sec.[147] Thus if it were possible to convert infrared to visible, even at relatively low efficiency, we might hope to have a more sensitive infrared detector than is currently available.

A second attractive feature of this method of detecting infrared is that the up-converter crystal, the laser pump (ω_2), and the photomultiplier all operate nominally at room temperature, 300°K, whereas the competitive detectors in the infrared commonly operate between 4.2 and 77°K.

Thus we conceive of our up-converter detector as a "box" containing a laser pump (ω_2) of as high a power as we can conveniently obtain, a mixing crystal, and a photomultiplier. The pump radiation is injected into the crystal along with some infrared that we wish to study. The two are mixed and, by means of suitable filters, are detected on the photomultiplier (Figure 6.1).

Our analysis of up-conversion is framed very much in these terms, emphasizing the features that are pertinent to the detection of infrared by frequency conversion to the visible. We loosely use the term "visible" to mean the range 4000 Å to 1 μ, which is approximately the sensitive range for photomultiplier cathodes but which exceeds the strictly visible range. We examine the characteristics of the overall conversion from infrared to visible

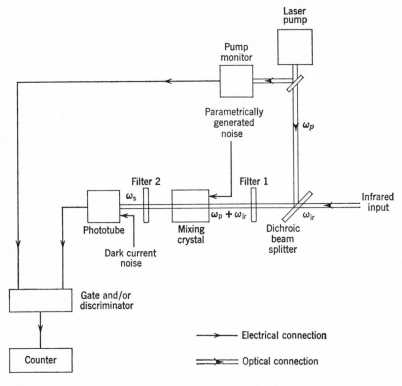

Figure 6.1 The schematic layout of a parametric up-converter for the detection of low-level infrared signals, including facilities for pulsed pumping. Also indicated are the noise-generating mechanisms within the system. Filter 1 passes only wavelengths equal to and longer than the pump wavelength, whereas filter 2 only passes wavelengths equal to the sum frequency or shorter. It may well be a line filter at the sum frequency.

and the noise characteristics of the detector, and we calculate some typical performance values. We also examine some of the implementation problems of these detectors and then turn our attention to another feature of the up-conversion process, namely, its ability to retain image information through the frequency-shifting process. This aspect has led to a considerable amount of interest in what has been termed the parametric image converter, and we examine it in some detail.

All the experimental and theoretical work that has been attempted in this field falls into two groupings. One approach has been to analyze and measure the conversion of a single mode of infrared radiation to the visible. This represents an extension of the case of SHG from a single transverse mode laser and has been treated in a similar manner. The alternative approach has been to consider the simultaneous conversion of a very large number of infrared modes, pumped by a single or multimode laser.

The former approach yields the most efficient conversion for that single mode, and in applications in which diffraction-limited optics are used for the collection of the radiation, it can yield very high conversion efficiencies on a continuous basis. The latter case yields a lower conversion efficiency per mode received, but this may be offset to some extent by the increase in the number of modes, so that the total signal power detected at the sum frequency is compensated. In addition, the latter case, through its ability to convert multi-mode infrared into multimode sum frequency in a way that allows retention of the mode relationships permits image information to be passed through the up-converter.

We examine both approaches, attempting to relate them to each other and to draw some conclusions on their present feasibility in different situations as competitive detectors of radiation.

6.2 GENERAL POINTS

We start by noting several features of equations 6.1 and 6.2. The time-averaged energy flux S is given by

$$S_i = \frac{cn_i}{8\pi} \mathscr{E}_i^{\;2} = N_i h\nu_i \tag{6.4}$$

where N is the photon flux. Using equation 6.4, we can recast equations 6.1 and 6.2 in the form

$$N_1(l) = N_1(0) \cos^2 \frac{l}{l_{puc}} \tag{6.5}$$

$$N_3(l) = N_1(0) \sin^2 \frac{l}{l_{puc}} \tag{6.6}$$

Adding these two relations together, we obtain directly

$$N_1(l) + N_3(l) = N_1(0) \tag{6.7}$$

which is the Manley–Rowe relation for the up-converter.

We can also show that

$$N_2(l) + N_3(l) = N_2(0) \tag{6.8}$$

We note two interesting properties of these relations. They show that the generation of one photon at the sum frequency ω_3 is accompanied by the annihilation of one quantum at the infrared (ω_1) and also one at the pump frequency (ω_2). Therefore, we are tempted to conclude that no sum frequency or visible quantum can be generated in the absence of any infrared quanta (i.e., there can be no spontaneous emission in the up-conversion process). In a detailed quantum-mechanical analysis, Louisell[103] has shown this assertion to be correct. This is in marked contrast to the parametric amplifier, where spontaneous emission is very significant. We return to both these points later.

Let us assume that, for the length of crystal used in our up-converter and the pump power that is available at ω_2, we have the condition $l \ll l_{puc}$. Now we can rewrite the foregoing equation, approximating sin (l/l_{puc}) to (l/l_{puc}) to obtain for unit area and plane waves the following relation (in c.g.s. units):

$$\frac{S_s(l)}{S_{ir}(0)} = \frac{N_s(l)\lambda_{ir}}{N_{ir}(0)\lambda_s} = \frac{512\pi^5 \, d^2 l^2 S_p(0)}{cn_{ir}n_p n_s \lambda_s^2} \tag{6.9}$$

or

$$\eta_{puc} = \frac{N_s(l)}{N_{ir}(0)} = \frac{512\pi^5 \, d^2 l^2 S_p(0)}{cn_{ir}n_p n_s \lambda_s \lambda_{ir}} \tag{6.10}$$

where η_{puc} is the quantum-conversion efficiency for the up-converter mixing stage and we have introduced the subscripts s, p, and ir to stand for sum, pump, and infrared in place of 3, 2, and 1. We shall see that the value of this conversion efficiency obtainable in practice is all-important in the development of competitive detectors.

The values obtainable are limited by the available combinations of material and pump laser that would allow useful interaction to occur. We note two restrictions placed on our choice. One is that the laser pump must be capable of high mean power operation; second, the sum frequency must be readily detectable. Finally, within these restrictions we must be able to phase match the interaction, that is, since $\omega_s = \omega_p + \omega_{IR}$, we must satisfy $k_s = k_p + k_{IR}$ in some convenient manner. It is this restriction that is usually the most serious.

6.3 FOCUSED BEAMS

In equations 6.9 and 6.10, it is clear that since S_p is the power per square centimeter, concentrating the available pump power into a smaller area will in general result in a greater conversion efficiency over that area. Thus for area A and pump power W_p (total—not per unit area), we have

$$\eta_{puc} = \frac{512\pi^5 \, d^2 l^2 W_p(0)}{cn_{ir}n_p n_s \lambda_{ir} \lambda_s A} \tag{6.11}$$

However, before we conclude from this that focusing increases our useful quantum efficiency, we should realize that, in the normal infrared detection situation, the field E_{ir} is derived from an incoherent source and its value is related to a brightness specified in watts per square centimeter per steradian per unit bandwidth. This means that the infrared power received generally decreases linearly with the area of the detector, so that the sum frequency total power remains constant. For this reason it is only in special cases that focusing gains anything in performance in the up-converter. Such a case would be the detection of TEM_{00} mode radiation from a laser source in the infrared. This case has been considered in detail by Kleinman and Boyd.[92]

This type of detection in the infrared does not at first appear to be very useful, since thermal sources are not usually considered in terms of modes. However, any diffraction-limited optical system imaging a single resolution point could be described in such terms, and in this context the up-converter has attracted some interest among infrared astronomers.

The analysis of Kleinman and Boyd is presented in terms of another potential application, namely, that of the detection of single-mode radiation carrying wide-band data along a light pipe. The outstanding conclusion of their analysis is that for such situations (just as in SHG from a single transverse mode laser source), there tends to be an optimum length of crystal and an optimum beam size for the most efficient conversion. The actual values corresponding to any particular situation are a function of the wavelengths, the refractive indices of the material, and the type of phase matching employed. The calculation of these values requires full knowledge of all the system parameters, together with the use of graphical data derived from a numerical evaluation of some of the expressions encountered in the analysis. However, for the simple case of single transverse mode beams, mixing collinearly with propagation in the x-y plane of a crystal such as lithium niobate, and with optimum focusing (beam sizes), Kleinman and Boyd's result reduces to the following simple expression for the quantum-conversion efficiency:

$$\eta_{sum} = \frac{2048\pi^5 \, d^2 LHP_p}{cn_p n_s \lambda_{ir}^2 \lambda_p} \tag{6.12}$$

where H is a numerical factor equal to 1.068.

This result also assumes that all the infrared power to be converted falls within the phase-matched bandwidth. If we insert into equation 6.12 a bandwidth that is inversely proportional to crystal length, as would be appropriate for considerations of power detection from broad-band sources, then the total detected sum frequency power becomes independent of crystal length.

Kleinman and Boyd's general results describe the conditions for the optimum detection of single transverse mode radiation. The earlier result (equation 6.11), which was derived for plane waves, implicitly describes the conversion efficiency obtained for multimode radiation arriving at the up-converter crystal within the phase-matched solid angle and bandwidth. Although the result was apparently derived for plane waves of single frequency, it can be used for multimode radiation because the infrared and pump waves are totally uncorrelated in phase and frequency, unlike the case of SHG. Therefore, the net power detected at the sum frequency in the presence of multiple modes of either or both infrared and laser pump can be described in terms of their powers as if they were single frequency. This can be proved by noting that in the summation for the sum-frequency power, the only terms that contribute to the intensity can be written together in the form

$$I_{sum} \sim \sum_i W_p^i \sum_j W_{ir}^j$$

where W^i is the power in the ith mode.

In order to compare the single-mode and multimode results, we must make some estimates of the number of allowed modes involved in the multimode case and then deduce the net signal power detected for a fixed amount of total pump power when the "detector" is looking at a given object or source with, alternatively, the single mode or the multimode collection optics attached and with the pumping beam geometry adjusted accordingly.

Just as the single-mode case represents a limiting one in which the system parameters are specified by diffraction theory, the extreme multimode case is also limiting—here the parameters are most readily described by geometrical optics, as illustrated in Figure 6.2. In the single-mode converter, the diffraction properties of the single-mode beam and the crystal define the pump beam diameter, the length of the crystal used, and the positioning of the focus. In the multimode case, the geometrical factors of phase-matched solid angle for infrared radiation and the illuminated area of the converter crystal define the number of modes received. In both instances, the phase-matched bandwidths are defined similarly. Thus in order to compare the two approaches, it is necessary to examine the relations defining these geometrical properties.

6.4 EFFECTS OF PHASE MATCHING

Tuning

We have already discussed the importance of satisfying the phase-matching requirement if we are to obtain efficient interaction in a traveling-wave interaction. The up-conversion process imposes its own special limitations. For the multimode converter, we are interested in a situation in which the

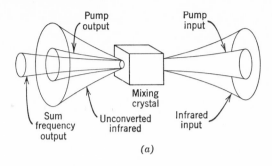

Pump output

Pump input

Mixing crystal

Sum frequency output

Unconverted infrared

Infrared input

(a)

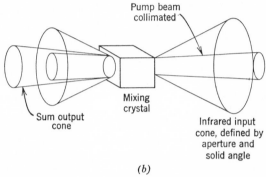

Pump beam collimated

Mixing crystal

Sum output cone

Infrared input cone, defined by aperture and solid angle

(b)

Figure 6.2 Schematic illustration of the difference between the single- and multiple-mode up-converters. (a) All the beams are single transverse mode and are focused for optimum conversion; (b) the infrared and sum cones contain many modes.

pump direction and frequency are fixed and well defined and the infrared source is diffuse and broad band. Thus neither infrared direction nor wavelength is precisely defined by the source, but is determined by the dependence of the conversion efficiency on the phase matching. Figure 6.3 presents the tuning curves for various types of up-converter as determined by phase matching. These curves are obtained by solving the phase-matching condition for fixed pump frequency and allowing the infrared frequency to vary. For the materials considered, the laser and infrared have been taken as ordinary rays and the sum as an extraordinary ray. In the case of lithium niobate, the waves are all considered to be propagating collinearly in the x-y plane, while the temperature of the crystal is varied to achieve different phase-matched wavelengths. For the proustite results, the waves are taken to be propagating collinearly but at some angle θ to the z axis, and the angle is varied in order to change the phase-matched wavelengths.

Thus in the conditions described, an infrared beam propagating in the laser beam direction, by virtue of a beam splitter introducing it, would be phase

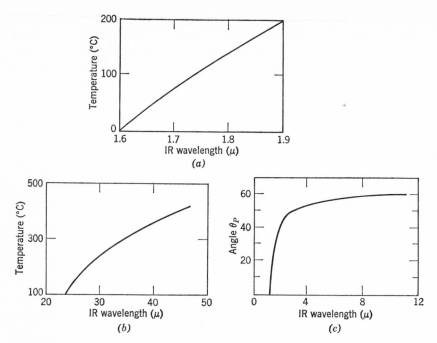

Figure 6.3 Tuning curves for some infrared to visible up-converters: (*a*) Ruby pump, lithium niobate; (*b*) 4880 Å argon, lithium niobate; (*c*) neodymium pump, proustite.

matched if its wavelength were that given by the curves of Figure 6.3. Infrared propagating at some other angle to the laser beam but entering the nonlinear crystal near to collinear would be phase matched at a slightly different set of wavelengths.

Frequency Bandwidth

Since the sum-frequency power as a function of phase mismatch varies as $\sin^2 (\Delta k \cdot L/2)/(\Delta k \cdot L/2)^2$, the effective bandwidth for phase-matched interaction is given by the relation

$$\Delta\omega = \frac{\pi}{2L\left(\dfrac{\partial k_{\mathrm{s}}}{\partial \omega} - \dfrac{\partial k_{\mathrm{ir}}}{\partial \omega}\right)} \tag{6.13}$$

where the differentials are evaluated at the frequencies appropriate for $\Delta k = 0$. This result gives the full bandwidth for the power to fall to $4/\pi^2$ on either side of the central maximum or to the 0.4 power points approximately.

In general, using equation 6.13, the bandwidth achieved for a 1-cm length of crystal L is very small, and since this bandwidth is tunable, the resulting detector can be considered as a narrow-band tunable detector. An experiment along these lines has been performed using lithium niobate as the non-linear crystal, tuning the frequency of the infrared converted from 1.6 to 2.4 μ.[112] The result (Figure 6.4) shows two traces of the output of a high-pressure

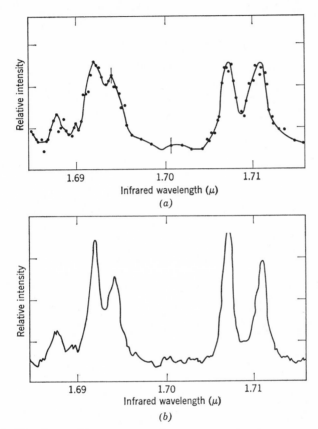

Figure 6.4 Use of a tunable up-converter detector to examine the mercury-arc spectrum: (a) The up-converter output using phase-matched tuning only; (b) the same spectrum using a cooled InSb detector with a double prism grating monochromator. After Midwinter and Warner, *J. Appl. Phys.*, **38**, 519 (1967).

mercury lamp, the upper taken with a conventional double monochromator with prism and grating, and the lower taken using only the finite phase-matched bandwidth of the up-converter to provide the spectral resolution. The crystal length was just over 1 cm and the resolution was of the order of 5 or 6 cm^{-1}. The resulting sum-frequency signal from the addition of 1.6 μ

and the ruby laser pump was in the region of 5000 Å and was thus readily detectable using a low-noise photomultiplier.

Although this fairly small bandwidth is typical of the up-converter, a much larger bandwidth can be achieved by design in special circumstances. This occurs when the material being used for the mixing process provides within its transmission band values of dispersion such that

$$\frac{\partial k_s}{\partial \omega} = \frac{\partial k_{ir}}{\partial \omega} \qquad (6.14)$$

With a particular pump laser equation 6.14 will be fulfilled only for a particular infrared wavelength. For example, when lithium niobate is pumped with a Nd:YAG laser, the condition is satisfied in the region of 3.5 μ; for the same material pumped by a ruby laser, the infrared wavelength is in the region of 5.5 μ. The pump wavelength is involved because, in order to satisfy the condition, it is necessary to place the sum and infrared frequencies on equal slope portions of the k versus ω curve, and the spacing of those frequencies depends on the pump frequency.

Using equation 6.14, an increase of bandwidth approaching 100 times the 5 to 6 cm^{-1} mentioned earlier has been demonstrated.[115] This was done in a Nd:YAG-pumped lithium niobate converter, in which the broad bandwidth occurred with a center wavelength of about 3.5 μ.

Solid Acceptance Angle for Infrared Radiation

As in the case of changing infrared frequency, phase-matching sets a limit to the solid angle in which a given infrared frequency can arrive in the crystal and still be up-converted. The phase mismatch is a function of the angle of incidence of the infrared; for efficient conversion, we take as before the condition that $|\Delta k| \leq \pi/L$. Figure 6.5 shows how this condition is related to the geometrical phase matching. Three cases are illustrated. In the first (Figure 6.5a), the cone of infrared radiation lies symmetrically about the pump radiation and both travel at an angle to the z axis of the crystal to achieve phase matching. The same is true in Figure 6.5b except that the angle is taken to be 90°, so that propagation is in the x–y plane of the crystal. Finally, Figure 6.5c shows propagation at a nominal angle to the z axis, but the infrared cone of directions is no longer symmetrically disposed about the pump direction. Each has its own special properties, and we discuss them in turn.

We assume here the normal phase-matching situation in a negative uniaxial crystal where the infrared and pump waves are polarized as ordinary rays and the sum as an extraordinary ray. Under these conditions, the first of the phase-matching geometries gives a relatively small acceptance angle in

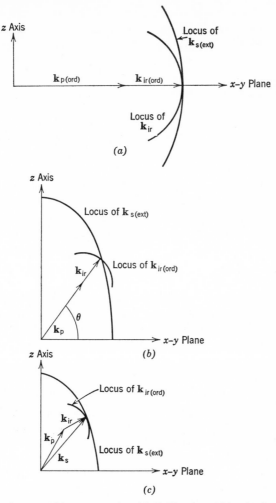

Figure 6.5 Three phase-matching geometries: (*a*) Collinear matching in the *x–y* plane of a crystal using temperature tuning, (*b*) collinear phase matching using angle tuning, (*c*) noncollinear phase matching using angle matching to obtain large acceptance angle.

the plane. As illustrated in Figure 6.6, this is because the surface defined by the vector $(\mathbf{k}_p + \mathbf{k}_{ir})$ marks out a spherical surface centered on the tip of $\mathbf{k}_p$ (assuming that the laser beam is well collimated, as is normally the case). However, the surface defined by the vector $\mathbf{k}_s$ is also a near-spherical one for small angles about the phase-matching direction, but the normal to that surface does not pass through the center of curvature of the $(\mathbf{k}_p + \mathbf{k}_{ir})$ surface because of the variation with angle of the index for the extraordinary

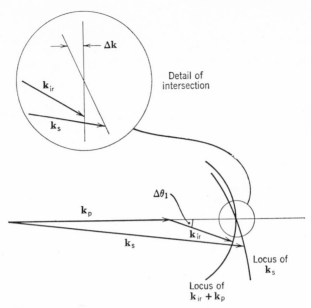

Figure 6.6 Detail of the angle-tuned collinear phase matching showing the allowed acceptance angle, in the angle-tuning plane.

ray. Thus the two surfaces cut each other at a small but finite angle and the change in angle of $\mathbf{k}_{ir}$ to produce the allowed Δk is small. Therefore the change in angle from the center that is allowed before a phase mismatch greater than Δk is produced is given by the simple expression

$$\Delta\theta_1 = \left(\frac{\partial k_s}{\partial\theta}\right)^{-1}\Delta k = \frac{\pi}{L}\left(\frac{\partial k_s}{\partial\theta}\right)^{-1} \tag{6.15}$$

We have assumed that the angles involved are small enough so that the spherical surfaces, in the section of the θ plane, can be considered to be linear.

In the direction perpendicular to the plane the surfaces of the vectors $(\mathbf{k}_p + \mathbf{k}_{ir})$ and $\mathbf{k}_s$ are both spherical and tangential to each other at the center of the beams, where perfect phase matching occurs. Since the two surfaces in this plane do not intersect asymmetrically, the form of the phase mismatch with angle is quite different (Figure 6.7). The half-angle $\Delta\theta_2$ is again the angle that produces a phase mismatch of π/L. By approximating the sine and the cosine of a small angle, we obtain the following result:

$$\Delta\theta_2 = \sqrt{2\pi/Lk_{ir}(1 - k_{ir}/k_s)} \tag{6.16}$$

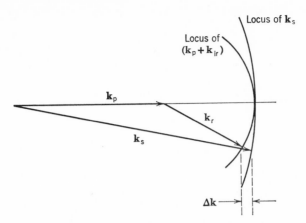

Figure 6.7 Detail of the angle-tuned collinear phase matching showing the allowed acceptance angle for the direction perpendicular to the angle-tuning plane.

The full solid angle for phase-matched radiation in the infrared, as measured external to the mixing crystal is then given by

$$\Delta\varphi_1 = 4n_{ir}^2\,\Delta\theta_1\,\Delta\theta_2 \tag{6.17}$$

and has an approximately rectangular cross setion. The factor n_{ir}^2 accounts for the angles derived in equations 6.15 and 6.16, which are measured inside the crystal and are, therefore, increased by a factor of approximately n_{ir} by refraction at the surface.

The second case involves collinear phase matching in which all the beams propagate with their centers lying in the crystalline x-y plane so that the effects of the extraordinary ray are not felt. In this case the surfaces $(\mathbf{k}_p + \mathbf{k}_{ir})$ and $\mathbf{k}_s$ are once again spherical, to good approximation, and are tangential to each other in the central direction. However, since the surfaces are now symmetrically disposed, it is advantageous to gain a little in total phase-matched solid angle by choosing the phase mismatch to be π/L in the beam centers (Figure 6.8). Under these conditions, and once again approximating the cosine and the sine of a small angle, we find that the half-angle $\Delta\theta_3$ is given by

$$\Delta\theta_3 = \sqrt{4\pi/Lk_{ir}(1 - k_{ir}/k_s)} \tag{6.18}$$

The full phase-matched solid angle is now given by the result

$$\Delta\varphi_2 = \pi(\Delta\theta_3 n_{ir})^2 \tag{6.19}$$

and forms a true cone. The resulting solid angle $\Delta\varphi_2$ is now considerably larger than that obtained in the previous situation $\Delta\varphi_1$. (We illustrate this with a numerical example shortly.)

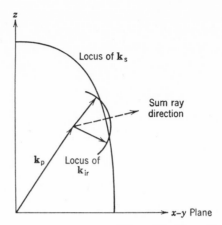

Figure 6.8 Noncollinear   angle-matching detail.

Finally, the third case, represented by Figure 6.9, is essentially the same as the second with regard to solid angle. However, it is of considerable device interest, since the method could be used to achieve the favorable phase matching of the second case, but without the requirement that the indices of the crystal used be tunable (thermally) so that exact phase matching occurs for the wavelengths of interest in the x-y plane. The method, proposed by Warner,[160] simply consists of setting the mean direction of the infrared cone of radiation along the ray direction rather than the wave-normal direction of the sum frequency. By this simple means, the surfaces of $(\mathbf{k}_{ir} + \mathbf{k}_{p})$ and $\mathbf{k}_{s}$ are made tangential at the center of the chosen cone of directions, and significant gains in phase-matched solid angles are obtained. The technique

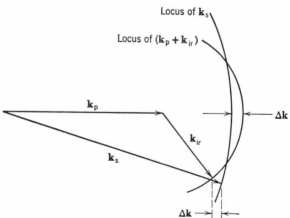

Figure 6.9 Temperature-tuned phase matching detail.

does not eliminate the problems associated with the linear displacement of the ordinary and extraordinary beams. As discussed in Chapter 5, for SHG this walk-off sets a definite limit to obtainable performance; however, in a multimode up-converter, it is of little consequence, since the beam diameters are typically large and the performance is governed by geometrical factors of solid angle, bandwidth, and aperture.

If we substitute numbers for a typical case in the foregoing expressions, the relative performances become clear. For the first situation we have $k_s = n_s \omega_s / c$, and thus

$$\frac{\partial k_s}{\partial \theta} = \frac{\omega_s}{c} \left(\frac{\partial n_s(\theta)}{\partial \theta} \right)_{\theta_m} \tag{6.20}$$

For an extraordinary ray, the differential is given by

$$\frac{\partial n_s(\theta)}{\partial \theta} = \frac{-n_s^e n_s^o [(n_s^o)^2 - (n_s^e)^2] \sin(2\theta_m)}{2[(n_s^o)^2 \sin^2(\theta_m) + (n_s^e)^2 \cos^2(\theta_m)]^{3/2}} \tag{6.21}$$

If, for example, we take $\theta_m = 45°$ and we use refractive index data typical of lithium niobate, with an infrared wavelength of 3.5 μ and sum wavelength of 0.5 μ, we find that the approximate results for a 1-cm crystal are:

$$\Delta\theta_1 = 1.7 \text{ mrad}$$
$$\Delta\theta_2 = 14.0 \text{ mrad}$$
$$\Delta\theta_3 = 19.0 \text{ mrad}$$
$$\Delta\varphi_1 = 440 \times 10^{-6} \text{ sr}$$
$$\Delta\varphi_2 = 5600 \times 10^{-6} \text{ sr} \tag{6.22}$$

In conclusion, we note from equation 6.19 that for x–y plane phase matching (Figures 6.8 and 6.9) or for noncollinear phase matching, the solid angles vary inversely with crystal length. For the critical case (equation 6.17; Figure 6.6), the solid angle varies as $L^{2/3}$ and is in general considerably smaller, typically by an order of magnitude or more. At the same time, the phase-matched bandwidth varies inversely as the crystal length and the conversion efficiency varies as the length squared. Hence for the best conditions (large solid angle), the amount of sum-frequency power generated from a broadband diffuse source is expected to be independent of the mixing crystal length. This is because the net infrared power varies as the product of the phase-matched solid angles and bandwidths.

6.5 COMPARISON OF THE SINGLE-MODE AND MULTIMODE UP-CONVERTERS

We are now in a position to compare directly the relative performances of the optimized single- and multimode up-converters. To do this we must first derive an expression for the number of modes received by the multimode up-converter.

The solid angle associated with a single transverse black body mode of the radiation field of wavelength λ is given by λ^2/A, where A is the area of the aperture from which it is emitted (or received). This result was proved generally by Yariv,[170] but it also follows from the analysis of Gaussian beams by Boyd and Kogelnik[25] (see Section 5.3).

By setting $z \gg b$ in equations 5.2 and 5.3, we find for the far-field beam divergence of a single Gaussian beam a half-angle

$$\Delta\theta = \frac{w_0'}{z} = \frac{\lambda}{\pi w_0} \tag{6.23}$$

Hence the solid angle is given by

$$\Delta\varphi = \pi \cdot \Delta\theta^2 = \frac{\lambda^2}{\pi w_0{}^2} = \frac{\lambda^2}{A} \tag{6.24}$$

Therefore, the number of modes per solid angle $d\Omega$ is $N_\Omega = (A/\lambda^2)\, d\Omega$. Now the number of modes in frequency space is easily derived from the number of modes in k space. The number of modes between k and $k + dk$ in a distance L is given by

$$N_\nu = \frac{L}{\pi}\,|dk| = \frac{2nL}{c}\,d\nu \tag{6.25}$$

For a radiation field in thermal equilibrium with blackbody surroundings, the number of quanta per mode is given by the relation[52]

$$q = \left[\exp\left(\frac{h\nu}{kt}\right) - 1 \right]^{-1} \tag{6.26}$$

Thus the number of quanta from a blackbody surrounding at $T^\circ K$, received in one second by an aperture of area A in a solid angle $d\Omega$, and a bandwidth $d\nu$ is given by

$$N_q = N_\Omega \frac{q}{2} N_\nu \tag{6.27}$$

where we have set $L = c/n$, the path traveled in one second. We use $q/2$ rather than q, since we seek the quanta received by an aperture rather than the quanta stored in the field, which can be coming or going! It also follows from the foregoing arguments that the number of modes received in a solid angle $d\Omega$ and bandwidth $d\nu$ by an aperture of area A is given by

$$N_m = \frac{A}{\lambda^2}\, d\Omega\, d\nu = \frac{A\nu^2}{c^2}\, d\Omega\, d\nu \tag{6.28}$$

when the radiation is centered on the frequency ν.

If we now take our result for the conversion efficiency for plane wave components within the solid angle $d\Omega$ and bandwidth dv with pumped aperture A (equation 6.11) and weigh it by the number of modes of infrared radiation that it receives from a diffuse source, we can compare it directly with the expression for the optimized conversion efficiency for a single mode of infrared. Thus for the multimode up-converter, the product of the number of transverse modes at frequency v_{IR} and the conversion efficiency per mode is given by equations 6.11, 6.19, and 6.28:

$$\eta_{\mathrm{puc}} N_\Omega \, \Delta\varphi_2 = \frac{1024\pi^6 \, d^2 L W_{\mathrm{p}}}{n_{\mathrm{s}} n_{\mathrm{p}} \lambda_{\mathrm{s}} \lambda_{\mathrm{ir}}{}^2 c[1 - n_{\mathrm{ir}}\lambda_{\mathrm{s}}/(n_{\mathrm{s}}\lambda_{\mathrm{ir}})]} \tag{6.29}$$

If we take the ratio of this result with the Kleinman and Boyd result in equation 6.12, we obtain

$$\frac{\eta_{\mathrm{puc}} N_\Omega \, \Delta\varphi_2}{\eta_{\mathrm{sum}}} = \frac{\pi\lambda_{\mathrm{p}}}{2\lambda_{\mathrm{s}} H[1 - n_{\mathrm{ir}}\lambda_{\mathrm{s}}/(n_{\mathrm{s}}\lambda_{\mathrm{ir}})]} \tag{6.30}$$

Since this factor is typically unity, we conclude that multimode operation is closely comparable to single-mode operation, to the accuracy of the approximations involved in this analysis.

Note that in both cases just cited, the conversion efficiencies obtained (equations 6.11 and 6.12) apparently increase with crystal length. But it must be remembered that this result neglects a circumstance that is evident from equation 6.13—namely, that the resolved bandwidth decreases with the length of the mixing crystal. Thus if a source has a linewidth greater than the phase-matched linewidth, nothing is gained in signal power by using longer mixing crystals in either case, although there will be a gain in spectral resolution. However, if a laserlike source with a narrow linewidth is to be detected, then a long mixing crystal would be advantageous.

We now calculate the quantum-conversion efficiencies for optimized single- and multimode up-converters for a specific case, to give the reader some feel for the meaning of the results derived. We take the equations already derived for multimode and single-mode conversion and adjust the multimode active area to just accommodate the appropriate number of modes of infrared radiation. We then evaluate the two equations using numbers representative of 3.5-μ infrared radiation and argon ion laser pump (5145 Å), a 1-cm long crystal of lithium niobate, and phase-matching of the type that leads to a large solid angle, $\Delta\varphi_2$. The results are presented in Figure 6.10. It is seen that even with one watt of laser pump power, high continuous-wave conversion efficiencies are possible for a small number of modes. By operating the same crystal inside the resonator of an argon ion laser, a circulating power of the

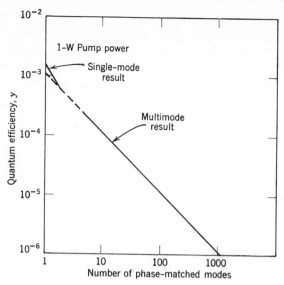

Figure 6.10 Plot of the theoretical conversion efficiency of an optimized up-converter based on a 1-cm crystal, typically lithium niobate; temperature-tuned phase matching and a pump power of about 1 W at 5000 Å are assumed.

order of 100 W or more should be available for pumping, which would lead to a corresponding increase in conversion efficiency.

6.6   NOISE PROPERTIES

Since the up-conversion process is nominally noiseless, in that it generates no signal at the sum frequency in the absence of any input, the signal counted by the phototube detector (assuming perfect filters in the optical system) is

$$N_{pm}{}^s = N_{ir}(L)\eta_{puc}\eta_{pm}$$

where $N_{pm}{}^s$ is the number of signal counts received and η_{pm} is the photo-cathode quantum efficiency. If $N_{pm}{}^n$ is the number of noise or dark-current counts received in the same time interval, the S/N ratio is given by

$$S/N = \frac{N_{pm}{}^s}{\sqrt{N_{pm}{}^s + N_{pm}{}^n}}$$

For a good photomultiplier, $N_{pm}{}^n$ is very small or zero. Hence the noise equivalent power of the detector (the value of N_{ir} per second for $S/N = 1$) is predominantly dependent on the quantum efficiencies of the conversion process and the photocathode. We may say, therefore, that the up-converter detector is characterized by relatively low quantum efficiency (Figure 6.10) and low noise, in marked contrast to the conventional photoconductive detector, which is characterized by high quantum efficiency (0.5) and high noise. This result has led to the prediction that the up-converter detector is most likely to find application in the area of very low-level infrared detection,[92] such as in spectroscopy or astronomy rather than in high-level, high-data-rate detection as in a communication link.

To see why this is so, consider two hypothetical but realistic examples. Consider an up-converter operating with a noise-equivalent power, (NEP) of 10^{-14} W and a conventional photoconductive detector at the same wave-l ngth with an NEP of 10^{-13} W. Assume that the wavelength of the infrared is approximately $3.5\,\mu$ so that the energy per quantum is 5.7×10^{-20} J. Then in order to produce a signal detectable above noise in one second, some 1.7×10^5 quanta must impinge upon the up-converter and some 1.7×10^6 quanta on the photoconductive detector. Either would give an S/N ratio of unity. Assuming that the photomultiplier used in our detector system has a quantum efficiency of 0.1, then the up-converter is operating at a quantum efficiency of 10^{-4} (assuming that $N_{pm}{}^n = 1$). Hence if the infrared signal power increases to 10^{10} quanta/sec, the upconverter S/N ratio will increase to about 300.

By comparison, since the photoconductor S/N ratio at low level is almost certainly set by background noise of one form or another (the equivalent of $N_{pm}{}^n$), and since its quantum efficiency is typically 0.5, we can by the same analysis calculate an effective value for $N_{pcd} = 72 \times 10^{10}$ so that, for a signal level in the infrared of 10^{10} quanta/sec, the S/N ratio becomes 10^4.

Thus, for low signal level detection, the critical measure of performance is typically NEP, but for signal input levels that are large compared with the NEP, the detector having the larger quantum efficiency is often better. The numbers for the foregoing cases are plotted in Figure 6.11 by way of illustration. The signal powers are taken to be mean power levels over the period of observation, in this case 1 sec. A more precise way of comparing detector performances in general situations uses information theory to calculate the amount of data that could be transmitted through a given element in a given time with a given amount of signal power and a specified or arbitrary means of coding. This approach was used by Kleinman and Boyd[92] in a discussion of up-converter and photoconductive detectors in which they considered the continuous detection of data at the end of a transmission line.

A further variable that has not been exploited in the up-converter is its ability to "store detection power." That is to say, since the quantum conversion efficiency depends on the laser pump power, and since most solid-state lasers are mean-power limited rather than peak-power limited by virtue of the energy storage provided by the inverted population, the up-converter

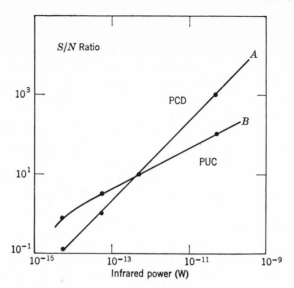

Figure 6.11 The signal to noise ratio for the two hypothetical detectors (PCD and PUC) as a function of infrared power, showing the superiority of the high-quantum-efficiency photoconductive detector (curve A), as the signal power increases, relative to the low-noise, low-quantum-efficiency parametric up-converter (curve B).

offers another possibility—namely, that of a detector that can operate at modest performance continuously or at very high performance in short bursts followed by periods in which it provides no detection capability. Such performance may be particularly suited to ranging applications in which the approximate time of return of a pulse is well known, or in certain types of synchronous detection related to pulsed sources of known cycle time and phase. At the time of writing, this synchronous pulsing capability of the up-converter has not been investigated experimentally, although a little analysis has been presented of the expected relative performance characteristics.[114]

We should also note that for very low signal levels, such as we have implicitly considered previously, the simple formulas presented are not entirely realistic. A more precise approach involves a consideration of the true statistical nature of the quanta involved using Poisson statistics. An analysis of the information capacity of quantum communications systems has been presented in a paper by J. P. Gordon.[69]

Finally, before leaving the subject of up-converter noise performance, we should point out that our assumption that no sum quanta can be generated in the absence of infrared quanta at the imput to the converter is not strictly correct, although it proves to be good in most cases. Tang[155] and independently Smith and Townes[145] have shown that the up-conversion process is not ideally noiseless but that parametrically generated noise at the infrared frequency could subsequently be up-converted in a two-step process. The nature of this noise-generating process is illustrated in Figure 6.12.

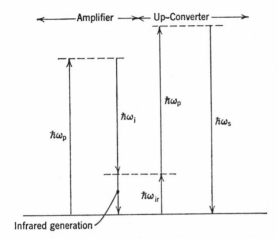

Figure 6.12 Schematization of the generation of (non-phase-matched) noise at the infrared frequency by the parametric amplifier in the same crystal as the phase-matched up-converter.

Infrared is spontaneously generated by the splitting of a pump quantum into an infrared quantum and an idler quantum. The infrared quantum is then up-converted to form a sum frequency quantum. The original parametric-generation process is not phase matched but nevertheless retains a finite probability.

The analysis of this noise process is complex; moreover, the formulas derived for the equivalent noise power at the sum or infrared frequencies depend on many variables, and several of them are critically dependent on the particular experimental situation.

Tang evaluated his results for several different sets of conditions. The one that appears to be most immediately relevant in contemporary up-conversion systems is one in which the noise-power bandwidth and solid angle are defined primarily by the detector (photomultiplier) and its optics and not by the up-conversion process. Then, if

$$\omega_s = \omega_p + \omega_{ir}$$
$$\omega_p = \omega_{ir} + \omega_d$$
$$\Delta k_{pa} = |k_p - k_{ir} - k_d|$$
$$\Delta k_{puc} = |k_s - k_p - k_{ir}|$$

Tang found that

$$W_{s(noise)} = \frac{16\, d_{pa}^2\, d_{puc}^2\, \hbar \omega_s^4 \omega_d \omega_{ir}^2 W_p^2 L^2 \Delta\omega_s\, \Delta\Omega_{det}}{\pi c^8 n_p^2 n_{ir}^2 n_s n_d\, \Delta k_{pa}^2 A}$$

where we have the following additional definitions:

d_{pa} = effective nonlinearity for the amplifier
d_{puc} = effective nonlinearity for the upconverter
L = crystal length
A = beam area
$\Delta\omega_s$ = phase-matched bandwidth at ω_s
$\Delta\Omega_{det}$ = phase-matched solid collection angle

We can recast Tang's result to give an equivalent noise power at the infrared by dividing by the power conversion efficiency. Two facts are worth noting:

1. The equivalent input noise power at the infrared due to the amplifier process depends linearly on the pump power. Since the equivalent noise power due to the ideal up-conversion process varies as the reciprocal pump power (i.e., the reciprocal of the quantum efficiency), it is clear that the two will cross at some point. Thus the parametric noise process ultimately becomes the limiting factor in detector performance.

2. Smith and Townes calculated that, for a ruby-laser-pumped lithium niobate up-converter operating at 1% quantum-conversion efficiency, the quantum noise contribution to the sum-frequency detected signal would be approximately 10^4 quanta/sec. If this conversion efficiency were maintainable on a continuous-wave basis, then the quantum noise so generated would be the limiting performance factor. However, in this example, the pump power was of the order of 10^6 W/cm² and the active area was 1 cm². Scaling the result to continuous-wave operating levels makes the effect generally negligible. The sole occasion when it might become troublesome appears to be when both the parametric up-conversion and amplification processes are nearly phase matched simultaneously, which can occur in rare situations. This normally requires that a significant amount of pump power be orthogonally polarized to the up-converter pump.

6.7 PARAMETRIC IMAGE CONVERTERS

Principles

Like all other optical nonlinear traveling-wave devices, the up-converter is constrained to satisfy the phase-matching condition

$$\Delta k = |\mathbf{k}_s - \mathbf{k}_p - \mathbf{k}_{ir}| \leq \frac{\pi}{L}$$

if efficient interaction of waves is to take place. In the image converter we exploit the vectorial nature of this relationship in a way that allows us to transfer image information from the infrared input to the sum frequency output. The technique for doing this was first outlined and demonstrated by Midwinter[113] and has been investigated further by several workers.

For simplicity we assume that our infrared object is located at infinity. In practice this would probably be done by the use of a spherical lens or mirror. Under these circumstances, the image information entering the up-converter is in the form of a series of plane waves with **k** direction specifying the original object point (Figure 6.13). That is to say, the spatial intensity

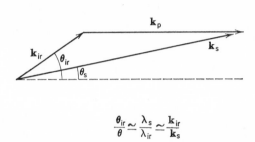

$$\frac{\theta_{ir}}{\theta} \sim \frac{\lambda_s}{\lambda_{ir}} \sim \frac{k_{ir}}{k_s}$$

Figure 6.13   Showing the coupling of angular information from an infrared input to the sum frequency output by mixing with a collimated pump beam in a parametric image converter.

information of the object undergoes a Fourier transformation into an angular distribution of information. Then if the laser pump beam is well defined and parallel (TEM$_{00}$ mode) so that its **k** vector is precisely located, angular information is transferred directly to the sum frequency.

Theoretically we can summarize this as follows. The image information in the infrared is a summation of plane waves in angular and frequency space

$$\mathscr{E}_{ir(image)} = \iiint \mathscr{E}_{ir}(\theta, \varphi, \omega_{ir})e^{i(\omega_{ir}t - \mathbf{k}_{ir}\cdot\mathbf{r})} \, d\theta \, d\varphi \, d\omega$$

The waves propagate fundamentally along the z direction with small departures given by θ and φ as in Figure 6.14. We also allow ω to cover a band of frequencies rather than a discrete line. Likewise, we may define our laser beam in similar terms although with discrete frequency:

$$\mathscr{E}_{p(total)} = \iint \mathscr{E}_{p}(\theta, \varphi)e^{i(\omega_p t - \mathbf{k}_p\cdot\mathbf{r})} \, d\theta \, d\varphi$$

In the image conversion system, a highly collimated laser beam is used, as explained previously. See Figure 6.15. Therefore, we approximate the term

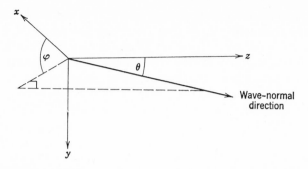

Figure 6.14 Definition of the coordinates used for the analysis of the image converter. The laser pump is directed along the z direction. After Midwinter, *IEEE J. Quant. Electr.*, **QE-4**, 717 (1968).

$\mathscr{E}_p(\theta, \varphi)$ to $\mathscr{E}_p \, \delta(\theta, \varphi)$: so that

$$\mathbf{k}_p = \begin{vmatrix} 0 \\ 0 \\ k_p \end{vmatrix}$$

The polarization wave produced in the nonlinear material by the interaction of laser and infrared radiation is now given by an expression of the form

$$\iiint \mathscr{E}_p \mathscr{E}_{ir}(\theta, \varphi, \omega_{ir}) e^{i(\omega_p + \omega_{ir})t - i(\mathbf{k}_{ir} + \mathbf{k}_p)\cdot \mathbf{r}} \, d\theta \, d\varphi \, d\omega$$

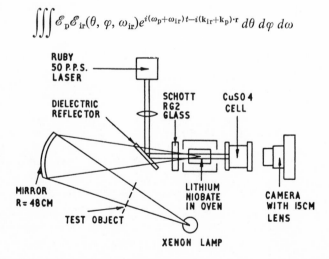

Figure 6.15 A typical experimental apparatus used for image-conversion studies. After Midwinter, *Appl. Phys. Lett.*, **12**, 68 (1968).

Then, assuming perfect phase matching, we may write the sum-frequency output in the form

$$\mathcal{E}_{\text{sum(image)}} = \iiint \mathcal{E}_s(\theta', \varphi', \omega_s) e^{i(\omega_s t - \mathbf{k}_s \cdot \mathbf{r})} \, d\theta' \, d\varphi' \, d\omega$$

where

$$\mathbf{k}_s = \begin{vmatrix} k_s \sin \theta' \cos \varphi' \\ k_s \sin \theta' \sin \varphi' \\ k_s \cos \theta' \end{vmatrix} \qquad \mathbf{k}_{ir} = \begin{vmatrix} k_{ir} \sin \theta \cos \varphi \\ k_{ir} \sin \theta \sin \varphi \\ k_{ir} \cos \theta \end{vmatrix}$$

and

$$\theta' = \varphi \qquad \text{and} \qquad \theta' \simeq \left(\frac{k_{ir}}{k_s}\right)\theta$$

This visible output is now of the same form as the infrared input; our image information, thus, has been conserved. In practice, the assumption that the laser beam will be a perfect plane wave traveling in the z direction is not satisfied, and this failure degrades the produced image.

Actually, the transformation introduces angular distortion when θ is large. In general, however, the distortion is probably not serious, since the phase-matching requirements limit the angular aperture.

The mode of operation just described is not the only one. Spherical rather than plane waves can be considered, and spatial information can be imaged onto the nonlinear material. However, these techniques introduce further distortion and limit the image quality obtainable. These approaches have formed the subject of detailed studies by R. A. Andrews[4] and, independently, A. H. Firester.[53]

Mode Analysis

An alternative approach to image converter analysis is to use the concept of modes that was considered earlier in this section. In this way we can easily derive some feel for the amount of picture information that can be handled by the up-converter. We have already calculated the number of modes of infrared radiation N_m that are received within the phase-matched solid angle and aperture of the up-converter (equation 6.28). These modes, being transverse modes, can be associated with picture information if we associate with each a discrete amplitude. Thus our N_m transverse modes of infrared can be associated with N_m discrete image points. If these are mixed with a single transverse mode laser, they will give rise to a sum-frequency beam containing N_m transverse modes. Thus we can use the already developed analysis for the multimode upconverter to describe the performance of the image converter. It then follows directly from equation 6.30 that, as the number of resolvable points is allowed to increase, the effective quantum conversion efficiency per data channel decreases in proportion.

This decrease should be compared with the equivalent situation for the photoconductive detector. In the simplest case, a single photoconductive detector serves to examine a single image point. The detector is then scanned over the scene in a TV raster-type scan to produce a picture on a cathode ray tube. Thus for N resolvable points, the detector spends $1/N$ of the total time examining each. Hence the features that may influence the choice of the up-converter in a given situation are factors such as the desirability of a directly visible image, the freedom from cooled detector and aperture elements, and performance in the particular system environment (rather than ability to obtain image information without sacrifice in other respects).

To obtain a general idea of the possible image-converter performance, we take the expression for the number of (blackbody) modes that can be received by the phase-matched converter, using the most favorable geometry. We find that it is given by the approximate relation already derived (equations 6.19 and 6.28), as follows:

$$N_m \, \Delta \varphi_2 = \frac{4\pi^2 n_{ir}^2 A}{L k_{ir} \lambda_{ir}^2 (1 - k_{ir}/k_s)} \simeq \frac{2\pi n_{ir} A}{L \lambda_{ir}(1 - \lambda_s/\lambda_{ir})}$$

If we insert numbers into this relation (using a 1-cm length of crystal, an infrared wavelength of 3.5 μ, and a refractive index of 2.0), then for a resolution, or number of discrete modes, equal to 10^5 (300 × 300 array), we find that the value of A obtained is typically of the order of 2 cm². To obtain an NEP of 10^{-11} W at 3.5 μ with an area of 2 cm² would typically require a pump power of several watts, arbitrarily allowing for 10% quantum efficiency in the photomultiplier. Thus this system would require a minimum power of 10^{-11} W at 3.5 μ *per image point* in order to produce a detectable signal over noise. To produce an image with some contrast would of course require more infrared power or pump power. As it stands, however, the detector behaves as a low-noise, low-quantum-efficiency array of 300 × 300 discrete elements, each element being operative all the time.

We can present these data in a different manner, beginning with the two sets of curves plotted in Figure 6.16. The dotted curves illustrate the number of quanta emitted per second, per mode, from blackbodies at 273°K in the regions of 10 and 3.5 μ. The number of quanta is plotted against the effective bandwidth in wavenumbers (cm⁻¹). Superimposed on these curves are the solid curves, which are the calculated numbers of quanta per mode required for a given S/N ratio using data typical of 1-W pump power and a 1-cm lithium niobate crystal. For each curve, the operating parameters have been optimized for a different number of up-converted modes (1, 100, and 10,000).

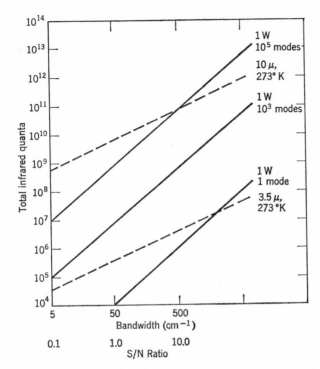

Figure 6.16 Curves showing the performance predictions for parametric image-converters as discussed in the text. The dotted curves represent the total infrared quanta emitted by a blackbody at room temperature as a function of bandwidth, centered at 3.5 and 10 μ, respectively, for a single transverse mode of the radiation field. The solid curves show the relation between the number of infrared quanta per received mode and the S/N ratio, for the hypothetical converter already discussed and illustrated in Figure 6.10. The three solid curves are, respectively, for converters optimized for 1, 10^3, and 10^5 received infrared modes.

Using the two sets of curves, and remembering that they share only the vertical axis (number of quanta), we can draw the following conclusions.

A single-mode converter with 5-cm^{-1} bandwidth should be readily capable of detecting the emission from a 273°K blackbody, with S/N ratios of about 8 when operating at 3.5 μ, and about 1000 when operating at about 10.0 μ. By contrast, an up-converter operating on 10^5 modes (a 300 × 300 array) and with 5-cm^{-1} bandwidth at 3.5 μ would show an S/N ratio of about 0.01 or less; at 10 μ the S/N ratio would be about 1.0. Thus substantial advances are required in the state of the art before we can hope to detect small temperature or emissivity differences in room-temperature objects using a parametric image-converter with TV resolution (300 × 300 array of points).

However, for an active imaging system operating at shorter wavelength, the approach might offer possibilities. Or, we may conclude that the practically attainable sensitivities for a very few image points make it attractive for a less sophisticated application than imaging in the normal TV picture sense.

6.8 EXPERIMENTAL STATUS OF UP-CONVERSION

The best experimentally measured value for NEP using an up-converter stands at 10^{-14} W in a single transverse mode at about 3.5 μ. This was obtained by Smith and Mahr[146] in an experiment employing lithium niobate as the mixing crystal and an argon ion laser to provide the continuous-wave pump radiation. The crystal was used outside the laser cavity, and its phase-matching temperature was high enough to overcome damage problems. This performance came close to that obtainable by conventional photoconductive detectors at the same wavelength. It is clear that this performance can, in principle, be improved by placing the mixing crystal inside the laser resonator, thus gaining one or two orders of magnitude in effective pump power. However, as we have already discussed for SHG, the considerable problems involved put an exceptional premium on the crystal quality, so that such experiments are not in general possible today.

In addition, efficient 10.6-μ conversion to the visible range has been studied by several workers,[30,56,93,161] using the single-mode optics and the multimode and imaging systems; but at this time no competitive detector systems have been built that operate at this wavelength. Imaging systems have been reported which offer excellent pictorial information, of the order of 100 to 300 lines resolution,[53,60,113] but this performance has always been achieved at the expense of sensitivity. Further developments in this area seem to await the discovery of more stable nonlinear materials that will allow the mixing crystal to be operated inside the laser cavity.

The evaluation of the up-converter as a viable detector of infrared radiation then requires careful study of its performance in a given system's environment, since, as we have seen in the foregoing discussion, its characteristics are a function of many design variables. Because of this it is most important to exercise great care in extrapolating experimental results obtained in one situation to a different situation.

7

Optical Parametric Amplification and Oscillation

7.1 INTRODUCTION

In Chapter 6 we pointed out that the parametric up-converter is a special case of sum-frequency generation. Similarly, the parametric amplifier and the parametric oscillator are special cases of difference-frequency generation. From the Manley–Rowe relations (Section 2.14) we know that in difference-frequency generation the photon with the highest frequency is split into two photons with lower frequencies: the energy taken away from the highest frequency beam is fed into the two beams of lower frequency. Consequently, the process can be used as an amplifier: a weak signal is made to interact with a strong, higher frequency pump, and both the generated difference frequency (known as the idler) and the original signal are amplified. If the idler and the amplified signal are passed through the mixing crystal again, with the proper phase, both are again amplified. In addition, if either one is passed through the crystal again with the proper phase, the result is still a gain in both. Thus the amplifier can be made into an oscillator by adding the proper feedback (i.e., a resonator) to *both* the signal and the idler, or by resonating only *one*. If the gain per path is larger than the loss per path, the signal can build up out of the noise, and the system will oscillate. If both the signal and the idler are resonated, the system of course has a lower threshold for oscillation than it would have if only one or the other were resonated. Other considerations may make this type of oscillator less preferable, however, as shown in Section 7.5.

Note that the denominations "signal" and "idler" have specific meanings only for the amplifier. In the oscillator, either of the two lower frequencies can be called the signal or the idler.

Amplifiers using the principles set out previously are well known at microwave frequencies.[101] Of particular interest to us is the varactor diode parametric amplifier because its operation closely resembles the process to be described here. The diode has the circuit characteristics of a voltage-dependent capacitance.[102] Thus the charge on the capacitor Q is given by a relation of the form $Q = C_1 V + C_2 V^2$. It follows directly from this that if a voltage of the form $V_1 \cos [\omega_1 t] + V_2 \cos [\omega_2 t]$ is applied across the capacitor, the charge will contain components varying at frequencies $\omega_1 + \omega_2$, $\omega_1 - \omega_2$, $2\omega_1$, etc. These can be used in the appropriate circuits to couple energy from the pump source into the signal and idler circuits. The great advantage of such an amplifier is that the elements of which it is composed can often be made nearly lossless, so that the amplifier can have an extremely low noise level.

The optical analog of the lossless electronic component is a perfectly transparent element. None of the radiation passing through such an element is absorbed, and so the element does not have to be in thermal equilibrium with the radiation. The thermal noise from a resistor in a microwave circuit thus finds its analog at optical frequencies in blackbody radiation, or thermal emission from any emissive body. However, as shown in Section 7.8, whereas thermal radiation can dominate the scene at microwave frequencies, this is rarely the case at optical frequencies because of the difference in the quantum energies involved.

The formal similarity between the expression just given for the charge stored on the varactor diode and the expression for the polarization in a nonlinear material is obvious, since $P = \alpha E + \beta E^2$, where P is the total polarization.

To derive the conditions and characteristics of the parametric amplifier at optical frequencies more precisely, we go back to the three coupled amplitude equations, equations 2.39.

In practice, parametric oscillators often operate with single transverse mode beams. However, the simple plane wave analysis, which we have also used for the description of SHG and up-conversion, demonstrates many of the features of the operation of a parametric oscillator, and because of its simplicity we use it as a starting point.

7.2 AMPLIFIER AND OSCILLATOR GAIN COEFFICIENTS

Since we expect to have a high power pump beam at frequency ω_3, the boundary condition to be used in equations 2.39 is

$$\frac{\partial E_3}{\partial z} = 0$$

Thus we have the two equations

$$\frac{\partial E_1}{\partial z} = -K_1 E_3 E_2^* e^{i\Delta kz} \tag{7.1}$$

$$\frac{\partial E_2}{\partial z} = -K_2 E_3 E_1^* e^{i\Delta kz} \tag{7.2}$$

where

$$K_j = \frac{8\pi i \omega_j^2 d}{k_j c^2}$$

By differentiating equation 7.1 once again with respect to z and using equation 7.2, we obtain a second-order differential equation, which for $\Delta k = 0$ has the general solution

$$\mathscr{E}_2(l) = \mathscr{E}_2(0) \cosh \frac{l}{l_{pa}} + \left(\frac{\omega_1^2 k_2}{\omega_2^2 k_1}\right)^{1/2} \mathscr{E}_1(0) \sinh \frac{l}{l_{pa}}$$

and the same for $\mathscr{E}_1$ by interchanging subscripts.

It is of interest to compare this solution with the solution for the parametric up-converter (equations 6.1 and 6.2), where we found a cosine and a sine dependence. It is easily seen that, mathematically, the difference in the solution exists because here we use the first and the second equations of equations 2.39, which both have a complex conjugate amplitude on the right-hand side, whereas for the up-converter we used the first and the third equations, the third having no complex conjugate amplitudes. The physical reason, of course, is that difference-frequency generation is a process that has gain, whereas sum-frequency generation does not.

The quantity l_{pa} is given by the relation

$$l_{pa} = \sqrt{n_1 n_2 \lambda_1 \lambda_2 / 8\pi d \, \mathscr{E}_3} \tag{7.3}$$

For large l the solution becomes

$$\mathscr{E}_2 = \left[\mathscr{E}_2(0) + \left(\frac{\omega_1^2 k_2}{\omega_2^2 k_1}\right)^{1/2} \mathscr{E}_1(0)\right] e^{(l/l_{pa})}$$

and we have an exponential gain coefficient $\alpha = 1/l_{pa}$. Thus we have two different sets of solutions. If l is of the same order of magnitude as l_{pa}, which is the case in the parametric amplifier, we have the solution

$$\mathscr{E}_2(l) = \mathscr{E}_2(0) \cosh(\alpha l) \qquad \text{signal} \tag{7.4}$$

$$\mathscr{E}_1(l) = \mathscr{E}_2(0) \left(\frac{\omega_1^2 k_2}{\omega_2^2 k_1}\right)^{1/2} \sinh(\alpha l) \qquad \text{idler} \tag{7.5}$$

whereas for the oscillator, where l is large, since the waves have transversed the crystal many times, we find solutions of the form

$$\mathscr{E}_2(l) = \mathscr{E}_2(0) e^{\alpha l} \tag{7.6}$$

$$\mathscr{E}_1(l) = \mathscr{E}_2(0) \left(\frac{\omega_1^2 k}{\omega_2^2 k_1}\right)^{1/2} e^{\alpha l} \tag{7.7}$$

It follows, that for $\alpha l < 1$, the amplifier wave builds up as $1 + [\alpha l]^2/2$ and the oscillator wave builds up as $1 + \alpha l$. The incremental power gains are therefore proportional to $[\alpha l]^2/2$ and αl, respectively, which for small αl means that the oscillator can have a gain large enough to overcome the roundtrip losses in a situation in which the single-pass amplifier shows only a negligible power gain.[28] Since αl is almost always small, the parametric amplifier has found little practical application at optical frequencies.

However, the result is of significance because an independent measurement of the gain of a potential parametric oscillator can be made by measuring the gain of the amplifier stage (i.e., the oscillator minus the resonator mirrors).

7.3 EFFECTS OF PHASE MISMATCH

Now let us examine the effects of a phase mismatch $\Delta k \neq 0$. By substituting in equations 7.1 and 7.2

$$E_1(z) = E_1(0)e^{sz+i\Delta kz/2}$$

and

$$E_2^*(z) = E_2^*(0)e^{sz-i\Delta kz/2}$$

we find

$$s^2 + \left(\frac{\Delta k}{2}\right)^2 = \frac{(8\pi d)^2\omega_1^2\omega_2^2}{c^4 k_1 k_2} E_3(0)E_3^*(0) \tag{7.8}$$

From equations 7.8 and 7.3 we find for s, the effective gain

$$s = \alpha - \frac{\Delta k}{2} \tag{7.9}$$

Thus we conclude that the presence of a momentum mismatch causes a reduction in the effective gain of the parametric amplifier.

It is tempting to conclude from the same result that the parametric amplifier has a gain-dependent bandwidth, since the phase mismatch is a function of the frequencies of the wave. However, care should be exercised in using the condition that the gain is zero for values of $\Delta k > 2\alpha$. This is because the foregoing analysis does not make proper allowance for the freedom of the parametric amplifier to choose its own phase for waves building up out of the noise. A more precise analysis shows that, when allowance is made for this effect, the gain expressions for an oscillator and an amplifier differ.[151]

The correct expressions under conditions of phase mismatch, analogous to the equations already derived for perfect phase matching, equations 7.4 through 7.7, are as follows. For the amplifier, converting to power, the result obtained is

$$S_2(l) = S_2(0)\left[1 + (\alpha l)^2 \frac{\sin^2(\Delta k l/2)}{(\Delta k l/2)^2} \right]$$

$$S_1(l) = S_1(0)(\alpha l)^2 \frac{\omega_1}{\omega_2} \frac{\sin^2(\Delta k l/2)}{(\Delta k l/2)^2}$$

Here we have assumed small gain, $\alpha l \ll 1$, which is normally the case for a continuous-wave optical parametric amplifier. The corresponding expressions for the oscillator, assuming that both signal and idler have built up from noise and that the loss rates for the two waves are equal, are

$$S_2(l) = S_2(0)\left[1 + 2\alpha l \frac{\sin(\Delta k l/2)}{\Delta k l/2} \right]$$

$$S_1(l) = \left(\frac{\omega_1}{\omega_2}\right) S_2(l)$$

where again the assumption is that $\alpha l \ll 1$.

From these results it is seen that simply inserting the adjusted gain derived in equation 7.9 into the expressions for the overall gain under perfect phase-matching conditions is not correct. For a detailed discussion of the effects of phase mismatch on the traveling-wave optical parametric amplifier, the reader is referred to the work of R. G. Smith,[151] who analyzed the mismatched parametric amplifier in general terms and examined the various cases of practical interest in detail.

7.4 PARAMETRIC OSCILLATION

We know that the condition for oscillation of the amplifier is simply that the gain must be larger than the loss. If the high Q resonator used to provide the proper feedback has mirrors with a reflectivity R, $[R \approx 1]$, then we require that $\alpha L \geq 1 - R$, where L is the length of the crystal. However, unless great care is taken, the losses will normally be larger than $1 - R$, due to imperfections of the crystal, and so on.

Optical parametric oscillation was first successfully demonstrated by Giordmaine and Miller,[66] who used a frequency-doubled, Q-switched CaWO$_4$:Nd laser to generate pump radiation at $\omega_p = 0.529 \mu$ (Figure 7.1).

The reflecting coatings that form the resonator for the idler and the signal were evaporated directly onto the flat and parallel ends of the lithium niobate oscillator crystal. These coatings were measured to have less than 0.4% transmission, but by measuring the finesse of the Fabry–Perot étalon formed by the crystal, the effective loss was determined to be 20%. The difference was ascribed to causes including crystal losses and scatter.

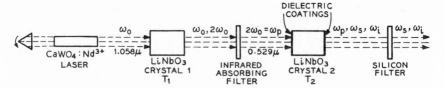

Figure 7.1 The parametric oscillator of Giordmaine and Miller. From Giordmaine and Miller, *Phys. Rev. Lett.*, **14**, 973 (1965).

Oscillation was observed with a pump power of 6.7 kW, which occurred in pulses 15 to 50 nsec long. This was calculated to correspond to an average pump intensity inside the resonator of 4×10^5 W/cm². The output of the oscillator was concentrated in a narrow spectral bandwidth and was tunable, due to the phase-matching requirement which was only satisfied over a narrow band of frequencies at any one time. Tuning was accomplished by thermally varying the lithium niobate refractive indices. The range of tuning obtained with this combination of pump and nonlinear crystal is illustrated in Figure 7.2, which shows that the range 0.7 to 2.0 μ was covered, representing a substantial band.

In these first experiments, the threshold power for oscillation was so high that only pulsed pump sources were powerful enough to produce parametric

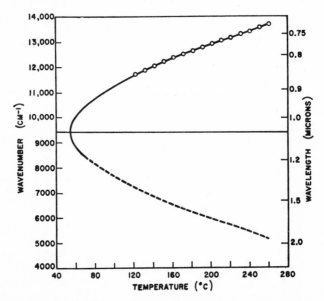

Figure 7.2 Tuning range of the oscillator of Figure 7.1. From Giordmaine and Miller, *Appl. Phys. Lett.*, **9**, 298 (1966).

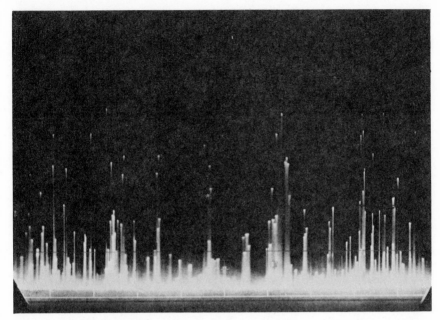

Figure 7.3   Spiking behavior of the output from the continuous-wave oscillator of Figure 7.4. From R. G. Smith et al., *Appl. Phys. Lett.*, **12**, 308 (1968).

oscillation. However, the experiments showed for the first time some of the difficulties encountered in attempts to obtain outputs that are stable in frequency. These problems are discussed in Section 7.5.

More recently Smith and co-workers[148] have demonstrated continuous parametric oscillation. They were able to achieve this as a result of several advances in technique over the work of Miller and Giordmaine. One major advance was the availability of the new nonlinear material $Ba_2Na_2Nb_5O_{15}$. Because the nonlinear coefficients of barium sodium niobate[143] are larger than those of lithium niobate, the new crystal allowed more efficient pumping. Smith also replaced the plane-parallel resonator used previously with a confocal resonator of higher Q value. The beams were focused in the resonator with carefully optimized parameters.

By varying the temperature of the crystal from 97 to 103°C, the output could be tuned from 0.98 to 1.16 μ. The spectral output was stable to within 5 Hz, although the temporal output consisted of a series of noiselike pulses (Figure 7.3).

With this combination, the continuous-wave threshold power was measured to be less than 45 mW, whereas 3 mW of oscillator power was seen from

300 mW of pump power. With later improvements, the same workers obtained a threshold as low as 3 mW of pump power.

The dramatic drop in threshold pump power came about because Smith et al. were able to use a TEM_{00} mode from a gas laser as the pump; they also coupled this mode very efficiently to the oscillator mode. Furthermore, the drop represented the culmination of an intensive effort to develop exceptionally high quality nonlinear materials. The optimum coupling of the TEM_{00} mode pump beam to the oscillator cavity is achieved by using a geometry similar to that already described for SHG and up-conversion from single transverse mode beams. Again, this type of coupling stems from a theoretical analysis by Boyd and Kleinman.[29]

The optimum coupling used was made possible by changing from an oscillator cavity formed by flat mirrors on the ends of the crystal to an essentially confocal resonator using curved mirrors. The actual geometry is clear from Figure 7.4. The resonator is formed from a curved external mirror

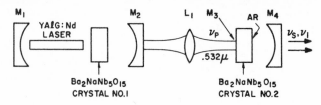

Figure 7.4 Schematic of the continuous-wave parametric oscillator of Smith. From R. G. Smith et al., *Appl. Phys. Lett.*, **12**, 308 (1968).

and a high-reflection coating applied to a flat face of the crystal. The other face of the crystal, lying inside the resonator, is antireflection coated. The resonator so formed is half of a normal confocal resonator but has the same desired feature of a very low diffraction loss for the lowest order mode. The resonator has the additional advantage, in this application, of presenting one less surface than there would have been had the resonator been made symmetrical. This is important, since any surface always imposes a loss, no matter how well finished or antireflection coated it is. The choice of mode size is governed by the condition for optimum energy transfer in the crystal of length L. If the neck of the mode is made too large, the power is insufficiently concentrated; if the neck is too small, the divergence at the exit faces of the crystal is too large for phase-matched interaction of the beams.

The results of Boyd and Kleinman's analysis for parametric generation are closely similar to their results for SHG and up-conversion, although each differs in matters of detail. The beam sizes for parametric generation are again described in terms of the confocal b parameter and are related by the relation $b_1 = b_2 = b_3 = b_0$. This is to a first approximation an experimentally convenient relation, since the idler and signal mirrors can then be the same surfaces. The presence of a dispersive crystal between them provides only a small perturbation.

For the more general situation, the choice of optimum beam size is a complex undertaking. However, for parametric generation in a crystal without double refraction (i.e., for phase matching in the x–y plane of a uniaxial material), the conditions simplify greatly and reduce to $b_0 = L/2.84$. Since the maximum in the gain versus beam size characteristic is rather flat, this condition can be taken rather loosely. It then follows that, for operation near degeneracy ($\omega_1 \approx \omega_2$), the gain coefficient reduces to the following form, which may be compared to the plane-wave result:

$$\alpha_{opt} = \frac{16\pi^3 d}{2.84\lambda_3} \sqrt{\pi W_3(0)/n_2 n_3 c b_0 \lambda_1}$$

Substituting numbers in this result, we find that for 1 W of pump power at 5000 Å, and using a 1-cm length of lithium niobate, the gain is typically 8×10^{-2}. Thus to produce oscillation in a resonator having 1% loss would require a single-mode pump power of the order of 15 to 20 mW. Note that in practice lithium niobate suffers severe optical damage even at these power levels, unless the crystal can operate at a temperature of the order of 200°C (see Section 4.5).

Referring now to the 3-mW threshold achieved by Smith et al., the magnitude of the achievement is clearer. The nonlinear coefficients of barium sodium niobate are approximately twice as large as those of lithium niobate. Thus the loss in their oscillator cavity must have been well below 1% per pass, which would require an exceptional quality crystal and coatings.

7.5 MODE HOPPING AND THE CLUSTER EFFECT

Both the parametric oscillators described in the previous section are doubly resonant. The early experiments of Giordmaine and Miller[67,68] very clearly demonstrated a problem associated with such doubly resonant parametric oscillators which seriously impairs their frequency stability and their tunability. It is associated with the fact that a resonator allows only certain modes.

So far we have implicitly assumed that ω_1 and ω_2 are simultaneously resonant in the cavity. Unfortunately this is usually not so; instead, the cavity is resonant at two frequencies ω_{10} and ω_{20} such that $\omega_3 = \omega_{10} + \omega_{20} + \Delta\omega$. For the frequencies ω_1 and ω_2, which are slightly off resonance, the resonator has a lower Q (i.e., the losses are larger), and thus the threshold for oscillation is higher. On the other hand, ω_1 and ω_2 are determined by the phase-matching condition, and thus while for the two frequencies ω_{10} and ω_{20}, for which $\Delta\omega = 0$, the Q of the cavity is large, the nonlinear interactions may be less efficient for these frequencies because the phase-matching condition is not satisfied. The operating point of the oscillator falls somewhere between these two extremes. However, in general, the decrease in Q due to a phase mismatch at $\Delta\omega = 0$ is much less than the decrease in Q due to $\Delta\omega \neq 0$ at $\Delta k = 0$. Therefore, the actual operating point of the oscillator can be quite far removed from the operating point determined from phase-matching data (Figure 7.5). This means that it is difficult to predict the operating frequencies

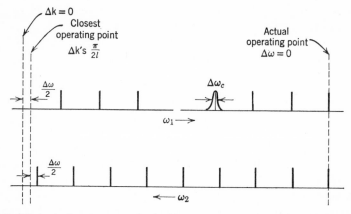

Figure 7.5 Modes in the resonator of a doubly resonant oscillator. Note that ω_1 increases to the right and ω_2 to the left. Lines vertically above each other indicate the coincidence of idler and signal modes. To the left is the point where ω_1 and ω_2 are phase matched. The closer operating point is the dotted line next to it, but the gain is higher at the point furthest to the right and this is the actual operating point. After Giordmaine and Miller, in *Physics of Quantum Electronics*, P. L. Kelley et al., Eds., McGraw-Hill, New York, 1966, p. 31.

of the oscillator. Moreover, since the operating frequencies are determined by the size of the resonator and the frequency of the pump, any small change in either quantity determines a different set of modes for which the gain is optimum, and thus the output shifts suddenly. This phenomenon is known as mode hopping. Since the gain is optimum for certain "clusters" of frequencies, the effect is often referred to as the cluster effect. The spiking behavior shown in Figure 7.3 is a typical result of mode hopping.

It can be shown that the effective gain s is given by[68]

$$s = -\alpha_{loss} \pm \frac{1}{2}\sqrt{\frac{4\pi^2\omega_1\omega_2\,d^2\mathscr{E}_3{}^2}{n^4}\left(\frac{\sin(\Delta kl/2)}{\Delta kl/2}\right)^2 - \Delta\omega^2} \qquad (7.10)$$

where $\Delta k = k_3 - k_{10} - k_{20}$ and $\Delta\omega = \omega_3 - \omega_{10} - \omega_{20}$, and $n = n_1 = n_2$.
Threshold occurs when $s = 0$. The threshold is lowest when $\Delta k = 0$ and
$\Delta\omega = 0$ (i.e., if the two phase-matched frequencies ω_1 and ω_2 "fit" exactly
in the resonator). If W_0 is the pump power required for this optimum case,
and the pump power required for $\Delta k \neq 0$ and $\Delta\omega \neq 0$ is W, then from
equation 7.10 we find

$$\frac{W}{W_0} = \left[1 + \left(\frac{\Delta\omega}{\Delta\omega_c}\right)^2\right] \bigg/ \frac{\sin^2(\Delta k l/2)}{(\Delta k l/2)^2} \qquad (7.11)$$

where $\Delta\omega_c = 2\alpha_{loss}$ is the full width of a longitudinal mode at half-intensity.
To calculate the magnitude of $\Delta\omega$, we first estimate the change in $\Delta\omega$
caused by increasing the mode number of both the signal and the idler waves
by unity. Allowing the mode number indices to change from m_1 to $m_1 + 1$
and from m_2 to $m_2 - 1$, the change in $\Delta\omega$ is given by

$$\frac{c\pi}{2n_{10}L} - \frac{c\pi}{2n_{20}L} = \frac{c\pi}{2L}\left(\frac{n_{20} - n_{10}}{n_{20}n_{10}}\right)$$

The maximum value that $\Delta\omega$ can take is just half the mode spacing, namely,
$c\pi/(4n_{10}L)$. Thus the maximum number of modes jumped in order to reach
a minimum value of the variable $\Delta\omega$ is given by the maximum value $c\pi/4n_{10}L$
divided by the change in $\Delta\omega$ for a unit mode jump. Hence the frequency shift
in jumping to a low value of $\Delta\omega$ is given typically by

$$\delta\omega = \frac{c\pi}{4n_{10}L}\left(\frac{n_{20}}{n_{20} - n_{10}}\right)$$

On the average, we may expect the oscillator frequency to shift by approximately half this amount in mode jumping. Nevertheless, the nearer the
degenerate oscillation is approached ($n_{20} = n_{10}$, $\omega_2 = \omega_1$), the more unstable the oscillator is likely to become. Even away from degeneracy, we find
that an evaluation of the factor ($n_2/n_2 - n_1$) for lithium niobate gives a
typical result in the range 10^1 to 10^3.

The extreme frequency instability around degenerate oscillation can be
partially overcome by using type II phase matching so that the two waves for
the signal and the idler are orthogonally polarized to each other. They then
see different refractive indices, one ordinary and one extraordinary, and the
frequency shift $\delta\omega$ is relatively small. This type of oscillator, demonstrated
by Akhmanov et al.,[2] may produce a greater frequency stability, but it offers
no gain in amplitude stability. In addition, interesting materials such as

lithium niobate do not allow type II phase matching, since the coupling goes to zero.

We should note that the build-up time for oscillation from noise is of the order of 0.1 to 1.0 microsec, since the continuous-wave oscillator is a low-gain device with high Q. It has been estimated that a mechanical stability of the order of 1×10^8 is required to obtain freedom from mechanical interference in the oscillator.

7.6 POWER LIMITING AND GAIN SATURATION

Siegman first showed that a parametric oscillator can act as an ideal power limiter in the sense that, once oscillation is achieved, the only pump power transmitted is the threshold power. He first demonstrated this (in collaboration with Ho) in a paper dealing with a microwave varactor–diode system;[75] subsequently, he considered the optical parametric oscillator.[141] We follow his analysis of the optical system.

For simplicity we consider degenerate oscillation with $\omega_s = \omega_i = \omega_p/2$ and make the following assumptions:

1. The interaction is phase matched.
2. The nonlinear crystal has the same linear optical properties as free space and completely fills the resonator which has mirrors at $z = 0$ and at $z = L$.
3. For ω_s and ω_i the mirrors have a reflectivity $R \approx 1$, but for ω_p they are 100% transmitting.

The independent waves propagating in the resonator structure are defined in Figure 7.6.

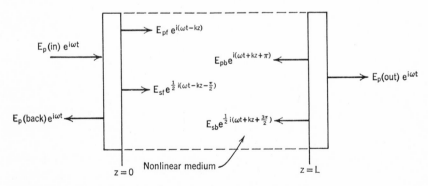

Figure 7.6   Waves in the doubly resonant parametric oscillator.

We write

$$\frac{\partial E_{pf}}{\partial z} = -KE_{sf}{}^2$$

$$\frac{\partial E_{pb}}{\partial z} = -KE_{sb}{}^2$$

$$\frac{\partial E_{sf}}{\partial z} = KE_{pf}E_{sf}^*$$

$$\frac{\partial E_{sb}}{\partial z} = KE_{pb}E_{sb}^*$$

where K is the nonlinear coupling coefficient. We look for steady-state solutions to these equations. Below threshold we have

$$E_{p(in)} = E_{pf} = E_{p(out)}$$

$$E_{p(back)} = E_{sf} = E_{sb} = E_{pb} = 0 \qquad (7.12)$$

In the steady state, the power gain in the signal waves must be just equal to their loss through the reflectors. Since $R \approx 1$, the amplitudes E_{sf} and E_{sb} are essentially constant throughout the crystal, and so we have

$$E_{p(out)} = E_{pf}(L)$$

$$= E_{pf}(0) - KE_{sf}{}^2L$$

$$= E_{p(in)} - KE_{sf}{}^2L \qquad (7.13)$$

and

$$E_{p(back)} = E_{pb}(0) = KE_{sb}{}^2L \approx KE_{sf}{}^2L \qquad (7.14)$$

Balancing net power input and net power output we obtain

$$E_{p(in)}^2 - E_{p(out)}^2 - E_{p(back)}^2 = 2(1-R)E_{sf}{}^2$$

and using equations 7.13 and 7.14

$$E_{p(in)}^2 - E_{p(out)}^2 - E_{p(back)}^2 = E_{p(in)}^2 - (E_{p(in)} - KE_{sf}{}^2L)^2 - (KE_{sf}{}^2L)^2 \qquad (7.15)$$

Equation 7.15 reduces to

$$E_{sf}{}^2 = \frac{1}{KL}\left(E_{p(in)} - \frac{1-R}{KL}\right) \qquad (7.16)$$

Substituting equation 7.13 in equation 7.16, we obtain

$$E_{p(out)} = \frac{1-R}{KL} \qquad (7.17)$$

From equation 7.16 we see that E_{sf} is only greater than zero if $E_{p(in)} > (1 - R)/KL$. Using this condition and equations 7.12 and 7.17, we see that

$$\frac{1 - R}{KL} = E_{th} \tag{7.18}$$

where E_{th} is the field at threshold. Thus we have

$$E_{p(out)} = E_{th}$$

We see then that the transmitted pump power is limited at the threshold power level. In other words, below threshold the transmitted pump power is equal to the incident pump power; but as the incident pump power is increased, the output pump power does not increase above the threshold power.

Where does the excess power go? It goes partly into the signal and idler outputs and partly back to the source in the reflected wave $E_{p(back)}$. To find out what fraction of the excess power goes back to the source, we calculate the power conversion efficiency

$$\frac{W_{s(out)}}{W_{p(in)}} = \frac{2(1 - R)E_{sf}^{2}}{E_{p(in)}^{2}}$$

and, using equations 7.16 and 7.18, we see that

$$\frac{W_{s(out)}}{W_{p(in)}} = \frac{2E_{th}(E_{p(in)} - E_{th})}{E_{p(in)}^{2}} \tag{7.19}$$

Equation 7.19 has a maximum value of 0.5 when $E_{p(in)} = 2E_{th}$.

It is convenient to express the different powers as functions of the ratio ρ between the pump power and the threshold power.

$$\frac{W_{p(back)}}{W_{th}} = (\sqrt{\rho} - 1)^{2} \tag{7.20}$$

$$\frac{W_{s(out)}}{W_{th}} = 2(\sqrt{\rho} - 1) \tag{7.21}$$

From equation 7.19 we see that the best conversion occurs for $\rho = 4$, but the optimum power conversion into signal and idler combined is only 50%. From equations 7.20 and 7.21 we learn that for $\rho < 9$ the larger fraction of the power above threshold is converted into signal and idler, whereas for $\rho > 9$ a larger fraction is reflected back to the source.

Siegman used his analysis to point out that the parametric oscillator can serve as a power limiter, but his conclusions that the maximum efficiency is 50% and that at this optimum point 25% of the input power goes back to the pump source are perhaps more important because of their significance to the operation of parametric oscillators to be used as tunable sources. The power going back to the source can seriously affect the stability of the pump and, consequently, the stability of the oscillator. This is because the resonator of the pump laser in effect sees the reflected power as a change in the reflectivity of its output mirror. The change is an increase or a decrease, depending on the exact optical distance between the oscillator and the pump. Thus air currents and other small variations can cause large instabilities.*

It is easily seen that this reflected beam is caused by the backward traveling signal and idler waves, which are always present in the nonlinear material. These waves regenerate a wave at the pump frequency which appears as reflected power.

It is of interest then to examine the maximum efficiency that can be achieved if the backward waves are eliminated. This can be done by using a ring resonator (Figure 7.7).

RING CAVITY PARAMETRIC OSCILLATOR

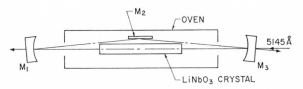

M_1 AND M_3: 5 cm RADIUS DIELECTRIC MIRRORS
M_2 : FLAT GOLD MIRROR

Figure 7.7 Ring resonant oscillator. After Byer et al., *Appl. Phys. Lett.*, **15**, 136 (1969).

With no backward traveling waves we have a greatly simplified set of equations

$$E_{p(in)} = E_p(0)$$

$$E_{p(out)} = E_p(L)$$

$$\frac{\partial E_p}{\partial z} = K E_s^2$$

$$\frac{\partial E_s}{\partial z} = K E_p E_s^*$$

* This effect can readily be shown by reflecting a fraction of the output of a helium–neon laser back into the laser—the laser output becomes extremely unstable, and the laser oscillation may stop altogether.

For steady-state oscillation we have again

$$E_p(L) = E_p(0) - KE_s^2 L$$
$$= E_{p(in)} - KE_s^2 L$$
$$= E_{p(out)}$$

and, equating the power in and out we have

$$E_{p(in)}^2 - E_{p(out)}^2 = 2(1 - R)E_s^2 \qquad (7.22)$$

so that

$$E_p^2(0) - (E_p(0) - KE_s^2 L)^2 = 2(1 - R)E_s^2$$

Thus we obtain

$$E_s^2 = \frac{2}{KL}\left(E_p(0) - \frac{1 - R}{KL}\right) \qquad (7.23)$$

so that

$$E_{th} = \frac{1 - R}{KL} \qquad (7.24)$$

Combining equations 7.22 to 7.24, we have

$$E_{p(out)} = 2E_{th} - E_{p(in)}$$

We find for the oscillator efficiency

$$\frac{W_{s(out)}}{W_{p(in)}} = \frac{4E_{th}(E_{p(in)} - E_{th})}{E_{p(in)}^2}$$

and thus, using again the ratio ρ between pump power and threshold power

$$\frac{W_{s(out)}}{W_{p(in)}} = \frac{4(\sqrt{\rho} - 1)}{\rho} \qquad (7.25)$$

Equation 7.25 has a maximum of 1 when $\rho = 4$. In other words, all the input power can be converted to signal and idler power, and no power is returned to the pump, if the oscillator is pumped four times over threshold.

This form of parametric oscillator was demonstrated by Byer et al.[37] They showed 60% depletion of the pump power (limited by the available pump power) and observed a very good isolation between the oscillator and the pump. By operating the same components also in a linear resonator configuration, they proved that the use of the ring resonator does indeed improve the frequency stability of the pump. It should be noted, however, that the ring resonator still forms a doubly resonated parametric oscillator; therefore, it is extremely sensitive to small vibrations, temperature changes, and other outside influences. Its output exhibits spiking behavior very similar to that shown in Figure 7.3.

7.7 MORE STABLE CONFIGURATIONS

Two types of solutions to the stability problem appear to be possible. One is to attempt to stabilize the resonator and its environment, passively or actively. The other is to circumvent the effect by avoiding simultaneous resonance of both the idler and the signal.

The first technique presents severe experimental problems. Stable resonance in a single mode from a doubly resonant parametric amplifier necessitates the simultaneous satisfaction of two requirements: the phase-matching condition and the condition that the frequency be the exact sum of two cavity resonant modes. In general, this indicates a need for two variables, and both must be accurately controlled. Boyd and Ashkin[28] have shown that it is theoretically possible to satisfy this condition by the use of temperature and electric field (acting through the electrooptic effect). If it is further required that the stable frequency be at a given frequency and be continuously, albeit stepwise, tunable; then a third variable is required. Conceptually, this variable might be the laser pump frequency, tuned over the emission line profile. However, it need hardly be stressed that the task of building such a three-element servo would be considerable.

An interesting solution of the second type is the backward wave oscillator proposed by Harris.[9,10,71] Here the pump and idler waves travel forward and the signal goes back (Figure 7.8). For the collinear situation, the following

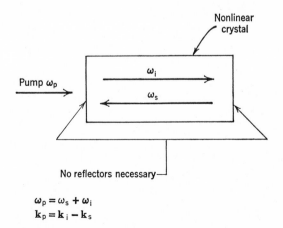

$$\omega_p = \omega_s + \omega_i$$
$$k_p = k_i - k_s$$

Figure 7.8 Backward wave oscillator.

conditions must satisfy

$$\omega_p = \omega_i + \omega_s$$
$$k_p = k_i - k_s$$
$$\omega_p > \omega_i > \omega_s \qquad (7.26)$$

Feedback is provided by the backward traveling wave. However, the practical problems of achieving the phase-matching condition (equation 7.26) have so far made the experimental demonstration of this type of parametric oscillator impossible.

Another solution is the parametric oscillator in which only the idler (or the signal, but not both) is resonant. The exact frequency of the idler is dictated by the phase-matching condition and is at most only one-half mode spacing away from the frequency that "fits" in the resonator. Thus the frequency stability is greatly increased, and the oscillator is more nearly continuously tunable.

The difference between singly and doubly resonant parametric oscillators has been shown in a series of experiments by Bjorkholm,[19-21] who used a single longitudinal mode Q-switched ruby laser to pump a crystal of lithium niobate. By changing the mirrors of the resonator in which the crystal is placed, he operated the oscillator in either the doubly resonant or the singly resonant configuration.

For his specific arrangement, Bjorkholm found that the tunability of the singly resonant oscillator was at least five times more accurate than the tunability of the doubly resonant oscillator. The operation of the singly resonant oscillator has been analyzed by Kreuzer,[94] who showed that, in the steady state, oscillation on a single actual mode should be possible. This was not observed by Bjorkholm, who pointed out that the steady state was most likely not achieved during his short pump pulses.

Bjorkholm also measured the incident and the transmitted pump power in both configurations and showed that in a doubly resonant oscillator, in the steady state, the transmitted pump power is indeed limited to the threshold power. His results for the singly resonant oscillator reveal no limiting action, and indeed they indicate zero pump power transmission at the point of efficient oscillation.

Figure 7.9 presents a commercially available parametric oscillator that uses a single resonant cavity. The oscillator appears as an integral part of the frequency-doubled Nd:YAG laser, which acts as its pump source. The laser can operate at several fundamental frequencies to yield a variety of pump frequencies. These are then used with different mirror combinations to provide almost complete coverage of the spectrum from 5000 Å to 4 μ. The nonlinear crystal used is lithium niobate, operated at high temperatures to avoid index damage problems.

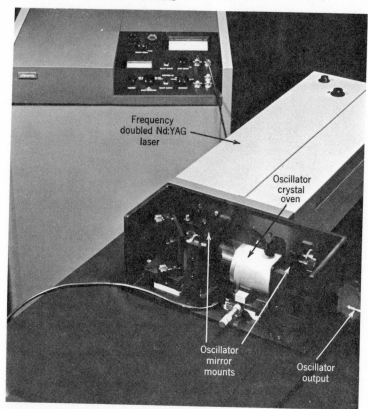

Figure 7.9 Chromatix Parametric Oscillator. Photo courtesy of Chromatix Inc., 1145 Terra Bella Boulevard, Mountain View, Calif. 94040.

A parametric generator in which no resonator at all is used on the amplifier crystal was demonstrated by Yarborough and Massey.[169a] As a pump they used the fourth harmonic of a Q-switched Nd:YAG laser, obtained by doubling the laser frequency in a 5-cm-long KD*P crystal and doubling the second harmonic again in a 5-cm-long ADP crystal. The amplifier crystal was also a 5-cm-long ADP crystal (Figure 7.10). By varying the temperature of the amplifier crystal the tuning curve of Figure 7.11 was obtained. The average pump power was 30 mW at 2662 Å, with a peak power of about 200 kW. The pump power density was calculated to be about 50 MW/cm². Average signal powers of up to 10 mW were measured.

For obvious reasons, this type of amplifier does not have any of the problems of alignment associated with the resonant oscillators. On the other hand, the signal and idler frequencies at which it operates are determined by the phase-matching condition only. Since the crystal used is fairly long, the

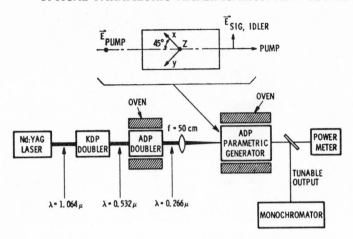

Figure 7.10 Nonresonant parametric generator. After Yarborough and Massey, *Appl. Phys. Lett.*, **18,** 438 (1971).

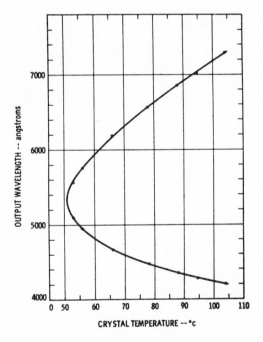

Figure 7.11 Tuning curve for the generator of Figure 7.10. After Yarborough and Massey, *Appl. Phys. Lett.*, **18,** 438 (1971).

condition limits the output to a relatively narrow frequency band; even so, the output bandwidth was measured to be 30 Å at degeneracy and 5 Å away from degeneracy. The point is somewhat moot, since many resonant oscillators do not perform much better, but it is mentioned to emphasize that the output bandwidth of a doubly resonant oscillator, properly controlled, is determined by the Q of the resonator, which is a more stringent condition than the phase-matching requirement.

The experiment amply shows the advantages of ADP-type crystals; namely, their resistance against optical damage and their availability in large, very good quality crystals: the gain coefficient in the amplifier crystal was calculated to be $\alpha = 2.7$, giving in the 5-cm-long crystal a power gain per pass of 1.4×10^{11}!

7.8 NOISE IN THE OPTICAL PARAMETRIC AMPLIFIER

We can distinguish two noise sources in the optical parametric amplifier; one is directly analogous to that observed in the microwave equivalent, and one is new. We shall see that it is the new one that predominates.

In the microwave amplifier, looking at a matched input at $T°K$, the thermal noise power received is kTB, where k is Boltzmann's constant and B the bandwidth. The optical equivalent to the matched input is a blackbody at $T°K$. Such a body radiates a power $W_{BB} = \hbar\omega B[\exp^{\hbar\omega/kT} - 1]^{-1}$ per single spatial mode where $\hbar = h/2\pi$, $h =$ Planck's constant. Notice that in the limit of $kT \gg \hbar\omega$, this reduces to $P_t = kTB$, which is the microwave case for room-temperature operation. (Here B is the number of frequency modes of the radiation field to which the amplifier responds.) If we evaluate the factor $(\hbar\omega/kT)$ for $300°K$ and the wavelength range 1 to 10 μ, we find that it varies from about 50 to 5. Hence, for the optical or infrared range, we are in the region $\hbar\omega \gg kT$, rather than vice-versa; this is what defines the noise performance.

The dominant source of noise in the optical parametric amplifier is no longer thermal in origin but stems from the $(\frac{1}{2})\hbar\omega$ zero-point energy per mode of the radiation field. The equivalent noise power input is $(\frac{1}{2})\hbar\omega_s B_s$ at the signal frequency, and $\frac{1}{2}\hbar\omega_i B_i$ at the idler frequency. Both contribute, and Kleinman[91] has shown that the noise contributions due to amplification of the signal and due to the mixing of the idler and the pump are equal and indistinguishable. It is, therefore, formally equivalent (but of no particular significance) to ascribe a noise input $\hbar\omega_s$ per mode at the frequency and to consider only the amplification term. The noise output power per mode is then $(e^{2\alpha l} - 1)\omega_s = 2\alpha l\hbar\omega_s B_s$ (for small gain), since we cannot detect the zero point energy. Figure 7.12 graphically presents the relative contribution

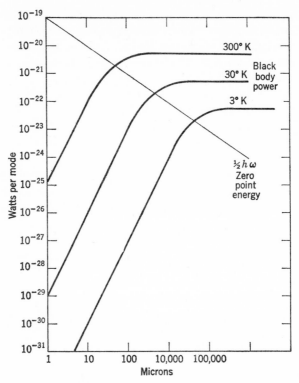

Figure 7.12   Relative contributions to the noise per mode from the zero-point fluctuation and from the blackbody terms.

to the noise per mode from the zero-point fluctuation and blackbody terms for a variety of temperatures and wavelengths.

The evaluation of the number of modes to which the amplifier responds, and the gain associated with them, depends critically on the experimental environment. Several authors have studied this problem, and we refer the reader to these papers.[36,64,73,91] Great care is required in the evaluation of B_s since collinear, noncollinear, and non-phase-matched processes can all contribute to the total noise.

7.9   REQUIREMENTS FOR THE LASER PUMP

For the doubly resonant oscillator, it is usually necessary for the laser pump to be a true single mode device if efficient operation is to be obtained. This is because the full gain is only achieved through the additive interaction of the modes at ω_p, ω_s, and ω_i. However, Harris[72] has shown that if the longitudinal mode spacing of the laser $\Delta\omega_p$ is identical to that of the doubly resonant oscillator cavity, then the gain at the signal or the idler frequencies is the same as it would have been, had the pump been single longitudinal mode. This is because all the pump and idler modes can interact with one signal mode of frequency ω_s, since

$$\nu_s = \sum_j (\nu_p + j\,\Delta\nu_p)(\nu - j\,\Delta\nu_i)$$

$$\Delta\nu_p = \Delta\nu_i$$

Normal dispersion usually prevents the more general condition from being satisfied, namely, that $\Delta\nu_i = \Delta\nu_s = \Delta\nu_p$. Fortunately such equivalence is not necessary. This form of pumping was used by Harris and co-workers[35] in the first demonstration of continuous-wave parametric oscillation, in which the pump was an argon ion laser operating in single transverse mode, but with many longitudinal modes. The oscillator cavity was approximately one meter long, to match the laser cavity, and the nonlinear crystal was lithium niobate.

For the singly resonant oscillator, true single-mode operation of the pump is not necessary. This is because the idler frequency is no longer constrained to be a cavity mode and can therefore adjust its linewidth to suit that of the pump. The allowed linewidth is then simply controlled by phase matching. Thus we find that the pump linewidth is given by

$$\Delta k = \left(\frac{\partial k_p}{\partial\omega} - \frac{\partial k_i}{\partial\omega}\right)\Delta\omega_p$$

or

$$\Delta\omega_p = \frac{\pi}{L\left(\dfrac{\partial k_p}{\partial\omega} - \dfrac{\partial k_i}{\partial\omega}\right)} \tag{7.27}$$

This is a much less stringent condition than the single longitudinal requirement. In practice it can allow the use of a pump linewidth of several wavenumbers, which might correspond to a 100 or more longitudinal modes. A particular example is given in Figure 7.13, which shows measurements made by Young et al.[171] The singly resonant oscillator used a 3.5-cm crystal of lithium niobate pumped at 0.473 μ and resonant at 2.79 μ to generate an idler at 0.569 μ. The measured linewidths are shown. This represents a favorable case, since the idler and the pump frequencies are close to each other, with the result that the values of the differentials in equation 7.27 will be similar, yielding a large allowed pump linewidth. If the resonant wave were made to be at 0.569 μ, the pump linewidth would need to be about one-sixth of the above mentioned amount.

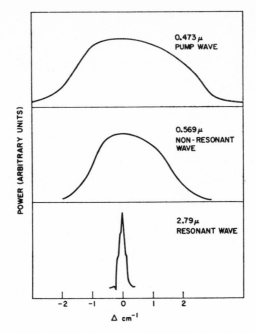

Figure 7.13   Curves showing that a narrow-band signal can be obtained in a singly resonant parametric oscillator pumped by a broadband laser. After Young et al., *J. Appl. Phys.*, **42**, 497 (1971).

The laser is usually operated in single transverse mode, since this yields the highest conversion efficiency or gain for a given amount of laser power when optimum focusing is employed. In addition, good amplitude and frequency stability are mandatory, if stability of the oscillator output is required. For this reason, the parametric oscillator probably presents the greatest challenge to builders of lasers and growers of nonlinear crystals alike, and will probably continue to do so for some time to come. For a more complete coverage of parametric oscillators, the reader is referred to a review paper by Harris.[73a]

Appendix 1

Tensors

A tensor is a physical quantity that relates a vector to one or more other vectors. If a vector $\mathbf{p}$ is dependent on a vector $\mathbf{q}$, each component of $\mathbf{p}$ will be dependent on all components of $\mathbf{q}$, and so we can write in terms of components

$$P_i = \sum_j T_{ij}q_j$$

The components T_{ij} form a second-rank tensor. Instead of writing the summation symbol, we employ the so-called Einstein convention and assume summation over repeated indices. Thus we write

$$P_i = T_{ij}q_j$$

If the vector $\mathbf{p}$ is dependent on two vectors $\mathbf{u}$ and $\mathbf{v}$ we can write

$$P_i = T_{ijk}u_j v_k$$

and the components T_{ijk} form a third-rank tensor. A vector can be considered as a first-rank tensor, and a scalar as a tensor of zero rank.

If the transformation from one set of rectangular coordinate axes to another is given by

$$x_i' = a_{ij}x_j$$

where x_1, x_2, x_3 are the coordinate axes, then the tensors transform according to

$$T_{ij}' = a_{ik}a_{jl}T_{kl} \tag{A1.1}$$

for second-rank tensors and to

$$T_{ijk}' = a_{il}a_{jm}a_{kn}T_{lmn} \tag{A1.2}$$

for third-rank tensors, and so on for higher ranks.

A tensor $[T_{ij}]$ is called symmetrical if $T_{ij} = T_{ji}$.

A symmetrical tensor $[T_{ij}]$ can be represented by a quadric that is defined as

$$T_{ij}x_i x_j = 1$$

This surface is a hyperboloid or an ellipsoid. By a suitable rotation of axes, it can be referred to its principal axes, whereupon it becomes

$$T_1 x_1{}^2 + T_2 x_2{}^2 + T_3 x_3{}^2 = 1$$

When referred to its principal axes, a symmetrical tensor $[T_{ij}]$ takes the form

$$\begin{vmatrix} T_1 & 0 & 0 \\ 0 & T_2 & 0 \\ 0 & 0 & T_3 \end{vmatrix}$$

and the linear relations $p_i = T_{ij}q_j$ are simplified to $p_1 = T_1 q_1$, $p_2 = T_2 q_2$, and $p_3 = T_3 q_3$.

When performing a sequence of axis rotations, care should be taken to observe the proper sequence in the tensor transformations. If the coordinate system is first rotated over an angle φ around the z axis, and then over angle θ around the new y axis, the resulting tensor becomes

$$\begin{vmatrix} \cos\theta & 0 & -\sin\theta \\ 0 & 1 & \cos \\ \sin\theta & 0 & \cos\theta \end{vmatrix} \begin{vmatrix} \cos\varphi & \sin\varphi & 0 \\ -\sin\varphi & \cos\varphi & 0 \\ 0 & 0 & 1 \end{vmatrix}$$

$$= \begin{vmatrix} \cos\theta\cos\varphi & \cos\theta\sin\varphi & -\sin\theta \\ -\sin\varphi & \cos\varphi & 0 \\ \sin\theta\cos\varphi & \sin\theta\sin\varphi & \cos\theta \end{vmatrix} \qquad (A1.3)$$

Note that the last rotation is the first matrix on the left-hand side. If the sequence of the two matrices on the left-hand side were reversed, we would obtain quite a different rotation. This is obvious, but the question of which sequence is the correct one can become quite confusing.

To illustrate the difficulty of determining the proper sequence, we take as an example the calculation of the indices and the directions of the allowed polarizations in a biaxial crystal for a ray at an angle θ to the z axis, and in a plane which makes an angle φ with the y axis. We assume the principal indices to be n_1, n_2, and n_3, and we write the indicatrix

$$\frac{x^2}{n_1{}^2} + \frac{y^2}{n_2{}^2} + \frac{z^2}{n_3{}^2} = 1$$

or, for brevity

$$a^2 x^2 + b^2 y^2 + c^2 z^2 = 1 \qquad (A1.4)$$

To find the required indices we rotate the coordinates over an angle φ around the z axis and an angle θ around the new y axis. The matrix for this

ʳotation is equation A1.3. However, to write the indicatrix in the new coordinate system we need the equations that give the old coordinates in terms
of the new ones; to obtain these, we have to perform the reverse rotation—
that is, we have to rotate the new coordinates x', y', and z' first over an angle
$-\theta$ around the y' axis and then over an angle $-\varphi$ around the z axis. This
rotation is given by

$$\begin{vmatrix} \cos\varphi\cos\theta & -\sin\varphi & \cos\varphi\sin\theta \\ \sin\varphi\cos\theta & \cos\varphi & \sin\varphi\sin\theta \\ -\sin\theta & 0 & \cos\theta \end{vmatrix} \qquad (A1.5)$$

and it gives

$$x = x'\cos\varphi\cos\theta - y'\sin\varphi + z'\cos\varphi\sin\theta \qquad (A1.6)$$

and so on for the other axes.

Substituting equation A1.6 in equation A1.4, and omitting the primes,
we find the following for the intersection of the indicatrix and the x', y'
plane:

$$[(a^2\cos^2\varphi + b^2\sin^2\varphi)\cos^2\theta + c^2\sin^2\theta]x^2 + (a^2\sin\varphi$$

$$+ b^2\cos^2\varphi)y^2 + [(b^2 - a^2)\cos\theta\sin 2\varphi]xy = 1 \quad (A1.7)$$

From equation A1.7 we see that the allowed directions of polarization
(the major and minor axes of the ellipse represented by the equation) are
parallel to the new coordinate axes only when $a^2 = b^2$; that is, when the
crystal is uniaxial.

Equation A1.7 can be referred to its principal axes by a rotation over an
angle γ around the new x axis where

$$\tan 2\gamma = \frac{(b^2 - a^2)\sin 2\varphi\cos\theta}{a^2(\sin^2\varphi - \cos^2\varphi\cos^2\theta) + b^2(\cos^2\varphi - \sin^2\varphi\cos^2\theta) - c^2\sin^2\theta}$$

The allowed polarization directions are parallel to the x and y axes obtained
after this last rotation, and the lengths of the major and minor axes of the
ellipse thus obtained are of course the required indices.

Instead of obtaining the old coordinates in terms of the new ones by re-
writing the sequence of the rotations, as we did, we can also make use of the
relation that

$$\text{if} \quad x'_i = a_{ik}x_k \quad \text{then} \quad x_i = a_{ki}x'_k$$

We see that this is true by comparing equations A1.3 and A1.5. For higher
rank tensors the law remains valid; for example, if $T'_{ijk} = a_{il}a_{jm}a_{kn}T_{lmn}$,
then

$$T_{ijk} = a_{li}a_{mj}a_{nk}T'_{lmn}$$

Appendix 2

Nonlinear Optical Susceptibilities

The nonlinear optical susceptibilities of a number of materials are listed in the Tables A2.1 through A2.9. Most of these coefficients were measured relative to a "known" crystal, as indicated in the tables. When no reference crystal is indicated, the values are given in 10^{-9} esu. In the visible range the "known" crystals are usually quartz, ADP, or KDP. In the infrared, at 10.6 μ, the "known" crystal is often gallium arsenide. The tables are grouped according to crystal classes, and the coefficients are listed as measured for second-harmonic generation; the listed wavelength is the fundamental. Only one value of the nonlinear susceptibility per wavelength is given for each material, even though more than one observation may have been made.

When comparing the merits of different materials, it should be remembered that the output of a frequency-mixing interaction is proportional to d^2/n^3. Since Miller's rule indicates that one of the requirements for a large d value is a large refractive index, the usefulness of a given nonlinear material cannot be determined from d alone. In some cases, therefore, a merit factor $M = d^2/n^3$ has been listed, as calculated using the refractive indices at the appropriate wavelengths. In addition, the efficiency depends on such factors as the beam walk-off.

The references are listed for each material and, where available, the reference giving index data is also listed.

For some materials the sign of the coefficient is specifically provided. These absolute signs are taken from the article by Miller and Nordland.[118]

For a more complete listing of many materials, the reader is referred to the article by R. Bechman and S. K. Kurtz in Landolt–Börnstein, *Numerical Data and Functional Relationships, Group III*, Volume 2, edited by K. H. Hellwege and published in Berlin by Springer-Verlag in 1969.

TABLE A2.1   VALUES FOR REFERENCE CRYSTALS

Quartz	.85
ADP	$1.36 \pm 12\%$
KDP	1.1
GaAs	320 ± 10

TABLE A2.2a   CLASS $\bar{4}$2m

Material	$\dfrac{d_{36}}{d_{36}(KDP)}$	M	λ	Ref.	Index Ref.
KH_2PO_4 (KDP)	1.0	.77	1.06	85	172
KD_2PO_4 (KD*P)	$.91 \pm .03$	.66	1.06	116	153
$NH_4H_2PO_4$ (ADP)[*]	$1.21 \pm .05$	.53	1.06	85	172
$NH_4D_2PO_4$ (AD*P)	1.10	.90	.694	154	
KH_2AsO_4	$1.12 \pm .05$		1.06	116	
CsH_2AsO_4	.53		.694	154	
RbH_2AsO_4	.64		.694	154	
RbH_2PO_4	1.04	.825	.694	154	153

[*] See Table A2.1 for the absolute value of ADP

TABLE A2.2b CLASS $\bar{4}$2m

Material	$\dfrac{d_{36}}{d_{36}(GaAs)}$	M	λ	Range	Ref.	Index Ref
ZnGeP$_2$	.83±15%	2430	10.6	.64-12	31	31
AgGaS$_2$*)	.134±15%	120	10.6	.5-12.5	33	33
CuGaS$_2$	.108±15%		10.6	.55-14.5	33	33
CuInS$_2$	.079±15%		10.6	.9-15	33	33
CdGeAs$_2$	3.4±20%		10.6	2.4-18	39	39

*)A value of .42±14 (d$_{14}$GaAs) is reported in 40

TABLE A2.3 CLASS $\bar{4}$3m

Material	d$_{14}$	λ	Reference	Index Ref.
InAs	1000±300	10.6	133	
CdTe	400±150	10.6	133	107
ZnS	.8 (d$_{33}$CdS)	1.06	152	
	73±20	10.6	133	
ZnSe	1.03 (d$_{33}$CdS)	1.06	152	107
	187±70	10.6	133	
ZnTe	3.47 (d$_{33}$CdS)	1.06	152	107
	220±80	1.06	133	
GaP	175±30 (d$_{36}$KDP)	1.06	116	
GaAs	560±40 (d$_{36}$KDP)	1.06	116	108
	320±100	10.6	109	
N$_4$(CH$_2$)$_6$	10 (d$_{36}$KDP)	1.06	74	

TABLE A2.4 CLASS 32

Material	d_{11}	λ	Ref.	Index Ref.
SiO_2	$.77 \pm .04$ (d KDP)	1.06	85,	139
Se	230 ± 60	10.6	47	59
Te	2200 ± 700	10.6	109	
$A\ell PO_4$	$.84 \pm .07$ (d_{36} KDP)	1.06	116	
HgS (Cinnabar)	150 ± 50 (d_{36} KDP)	10.6	83	83
$K_2 S_2 O_6$	$.24 \pm .04$ (d_{11} SiO_2)	.694	76	76
Benzil	11.2 ± 1.5 (d_{11} SiO_2)	1.06	86	

TABLE A2.5 CLASS 222

Material	d_{14}	λ	Ref.	Index Ref.
Ammonium oxalate $(NH_4)_2$ $C_2 O_4 H_2 O$	$.9-1.5$ (d KDP)	.6943	81	
Hippuric Acid $C_6 H_5 \cdot CO \cdot NH \cdot CH_2 CO_2 H$	≈ 5 (d_{36} ADP)	.6943	132	
HIO_3 α Iodic Acid	20 ± 5 (d_{11} SiO)	1.06	96	96

TABLE A2.6 CLASS mm2

Material	d_{15}	d_{24}	d_{33}	Ref. Crystal	λ	Ref.	Index Ref.
$Ba_2Na\ Nb_5O_{15}$	23 ± 7	20 ± 2	28 ± 2	$d_{11}(SiO_2)$	1.06	62	143
$LiCHO_2 \cdot H_2O$	$.3\pm.05$	$3.5\pm.25$	$5.1\pm.3$	$d_{11}(SiO_2)$	1.06	142	142
$LiGaO_2$	$+17\pm.02$	$\mp.37\pm.04$	$+1.45\pm1.5$	$d_{36}(KDP)$	1.06	119	

TABLE A2.7 CLASS 3m

Material	d_{31}	d_{22}	d_{33}	Ref. Crystal	λ	Ref.	Index Ref.
$LiNbO_3$	-10.6 ± 1.0	$+5.1\pm2.0$	-107 ± 20	$d_{36}(KDP)$	1.15	26	
$Ag_3As\ S_3$ Proustite	30	50		$d_{36}(KDP)$	1.15	80	80

TABLE A2.8 CLASS $\bar{4}$

Material	d_{31}	d_{36}	Ref. Crystal	λ	Ref.	Index Ref.
$CdHg(SCN)_4$	1.31	$0.3\pm.1$	$d_{33}(LiIO_3)$	1.06	14	14
$ZnHg(SCN)_4$						

TABLE A2.9 CLASSES 6mm, 4mm, AND 6

Material	Class	d_{15}	d_{31}	d_{33}	Ref. Crystal	λ	Ref.	Index Ref.
ZnO	6mm	4.7±.4	+4.3±4	-14.3±4	d_{36}(KDP)	1.06	116	
CdS	6mm	35±3	-32±2	+63±4	d_{36}(KDP)	1.06	152	
		69±17	63±15	105±30	absolute	10.6	133	
ZnS	6mm	51±20	45±15	.44	d_{36}(CdS)	1.06	152	
				89±30	absolute	10.6	133	
CdSe	6mm	74±18	68±15	2.59	d_{36}(CdS)	1.06	152	
				130±30	absolute	10.6	133	
BeO	6mm		-.46±.03	-.63±.04	d_{11}(SiO$_2$)	1.06	87	128
SiC	6mm	25±3	27±3	45±5	d_{11}(SiO$_2$)	1.06	144	144
LiIO$_3$	6		11.9±1.0	12.4±1.0	d_{36}(KDP)	1.06	121	
BaTiO$_3$	4mm	35±3	-37±3	-14±1	d_{36}(KDP)	1.06	116	

185

References

1. Adams, N. I., and Barrett, J. J., unpublished results.

2. Akhmanov, S. A., Korrigin, A. I., Kolosov, V. A., Piskarskas, A. S., Fadeev, V. V., and Khokhlov, R. V., "Tunable Parametric Light Generator with KDP Crystal," *Sov. Phys. J. Expt. Theor. Phys. Lett.* (Engl. transl.), **3**, 241 (1966).

3. Anderson, D. B., and Boyd, J. T., "Wideband CO_2 Laser Second Harmonic Generation Phase Matched in GaAs Thin Film Waveguides," *Appl. Phys. Lett.*, **19**, 266 (1971).

4. Andrews, R. A., "IR Image Parametric Up-Conversion," *IEEE J. Quant. Electr.*, **QE-6**, 68 (1970).

5. Armstrong, J. A., Bloembergen, N., Ducuing, J., and Pershan, P. S., "Interactions Between Light Waves in a Nonlinear Dielectric," *Phys. Rev.*, **127**, 1918 (1962).

6. Armstrong, J. A., "Measurement of Picosecond Laser Pulse Widths," *Appl. Phys. Lett.* **10**, 16 (1967).

7. Ashkin, A., Boyd, G. D., Dziedzic, J. M., Smith, R. G., Ballman, A. A., Levinstein, J. J., and Nassau, K., "Optically Induced Refractive Index Inhomogeneities in $LiNbO_3$ and $LiTaO_3$," *Appl. Phys. Lett.*, **9**, 72 (1966).

8. Ashkin, A., Boyd, G. D., and Kleinman, D. A., Phasematched Second Harmonic Generation without Double Refraction," *Appl. Phys. Lett.*, **6**, 179 (1965).

9. Aslaksen, E. W., "Threshold of the Optical Backward Wave Parametric Oscillator," *IEEE J. Quant. Electr.*, **QE-6**, 612 (1970).

10. Aslaksen, E. W., "The Optical Backward Wave Parametric Oscillator Above Threshold," *Opt. Commun.*, **2**, 69 (1970).

11. Bardsley, W., Davies, P. H., Hobden, M. V., Hulme, K. F., Jones, O., Pomeroy, W., and Warner, J., "Synthetic Proustite (Ag_3AsS_3): A Summary of Its Properties and Uses," *Opto-Electronics*, **1**, 29 (1969).

12. Bass, M., Franken, P. A., Ward, J. F., and Weinreich, G., "Optical Rectification," *Phys. Rev. Lett.*, **9**, 446 (1962).

13. Bergman, J. G., Ashkin, A., Ballman, A. A., Dziedzic, J. M., Levinstein, H. J., and Smith, R. G., "Curie Temperature, Birefringence, and Phasematching Temperature Variations in $LiNbO_3$ as a Function of Melt Stoichiometry," *Appl. Phys. Lett.*, **12**, 92 (1968).

14. Bergman, J. G., McFee, J. H., Crane, G. R., "Nonlinear Optical Properties of $CdHg(SCN)_4$ and $ZnHg(SCN)_4$," *Mat. Res. Bull.*, **5**, 913 (1970).

15. Bey, P. P., and Rabin, H., "Coupled Wave Solution of Harmonic Generation in an Optically Active Medium," *Phys. Rev.*, **162**, 794 (1967).

16. Billings, B. H., "The Electro-Optic Effect in Crystals of the Type XH_2PO_4," *J. Opt. Soc. Am.*, **39**, 797 and 802 (1949).

17. Bjorkholm, J. E., and Siegman, A. E., "Accurate CW Measurements of Optical Second Harmonic Generation in Ammonium Dihydrogen Phosphate and Calcite," *Phys. Rev.*, **154**, 851 (1967).

18. Bjorkholm, J. E., "Relative Signs of the Optical Nonlinear Coefficients d_{31} and d_{22} in $LiNbO_3$," *Appl. Phys. Lett.*, **13**, 36 (1968).

19. Bjorkholm, J. E., "Efficient Optical Parametric Oscillation Using Doubly and Singly Resonant Cavities," *Appl. Phys. Lett.*, **13**, 53 (1968).

20. Bjorkholm, J. E., "Some Spectral Properties of Doubly and Singly Resonant Pulsed Optical Parametric Oscillators," *Appl. Phys. Lett.*, **13**, 399 (1968).

21. Bjorkholm, J. E., "Some Effects of Spatially Non-Uniform Pumping in Pulsed Optical Parametric Oscillators," *IEEE J. Quant. Electr.*, **QE-7**, 109 (1971).

22. Bloembergen, N., and Pershan, P. S., "Lightwaves at the Boundary of Nonlinear Media," *Phys. Rev.*, **128**, 606 (1962).

23. Bloembergen, N., *Nonlinear Optics*, New York: Benjamin, 1965.

24. Bloembergen, N., and Sievers, A. J., "Nonlinear Optical Properties of Periodic Laminar Structures," *Appl. Phys. Lett.*, **17**, 483 (1970).

25. Boyd, G. D., and Kogelnik, H., "Generalized Confocal Resonator Theory," *Bell Syst. Tech. J.*, **41**, 1347 (1962).

26. Boyd, G. D., Miller, R. C., Nassau, K., Bond, W. L., and Savage, A., "$LiNbO_3$: An Efficient Phase Matchable Nonlinear Optical Material," *Appl. Phys. Lett.*, **5**, 234 (1964).

27. Boyd, G. D., and Patel, C. K. N., "Enhancement of Optical Second Harmonic Generation (SHG) by Reflection Phase Matching in ZnS and GaAs," *Appl. Phys. Lett.*, **8**, 313 (1966).

28. Boyd, G. D., and Ashkin, A., "Theory of Parametric Oscillator Threshold with Single Mode Optical Masers and Observation of Amplification in $LiNbO_3$," *Phys. Rev.*, **146**, 187 (1966).

29. Boyd, G. D., and Kleinman, D. A., "Parametric Interaction of Focused Gaussian Light Beams," *J. Appl. Phys.*, **39**, 3597 (1968).

30. Boyd, G. D., Bridges, T. J., and Burkhardt, E. G., "Up-Conversion of 10.6 μ Radiation to the Visible and Second Harmonic Generation in HgS," *IEEE J. Quant. Electr.*, **QE-4**, 515 (1968).

31. Boyd, G. D., Buehler, E., and Storz, F. G., "Linear and Nonlinear Optical Properties of $ZnGeP_2$ and CdSe," *Appl. Phys. Lett.*, **18**, 301 (1971).

32. Boyd, G. D., Gandrud, W. B., and Buehler, E., "Phase-Matched Up-Conversion of 10.6-μ Radiation in $ZnGeP_2$," *Appl. Phys. Lett.*, **18**, 446 (1971).

33. Boyd, G. D., Kasper, H., McFee, J. H., "Linear and Nonlinear Optical Properties of $AgGaS_2$, $CuGaS_2$ and $CuInS_2$ and Theory of the Wedge Technique for the Measurement of Nonlinear Coefficients," *IEEE J. Quant. Electr.*, **QE-7**, 563, (1971).

34. Bridenbaugh, P. M., Carruthers, J. M., Dziedzic, J. M., and Nash, F. R., "Spatially Uniform and Alterable SHG Phase-Matching Temperatures in Lithium Niobate," *Appl. Phys. Lett.*, **17**, 104 (1970).

35. Byer, R. L., Oshman, M. K., Young, J. F., and Harris, S. E., "Visible CW Parametric Oscillator," *Appl. Phys. Lett.*, **13**, 109 (1968).

36. Byer, R. L., and Harris, S. E., "Power and Bandwidth of Spontaneous Parametric Emission," *Phys. Rev.*, **168**, 1064 (1968).

37. Byer, R. L., Korrigin, A., and Young, J. F., "A CW Ring-Cavity Parametric Oscillator," *Appl. Phys. Lett.*, **15**, 136 (1969).

38. Byer, R. L., Young, J. F., and Feigelson, R. S., "Growth of High-Quality LiNbO$_3$ Crystals from the Congruent Melt," *J. Appl. Phys.*, **41**, 2320 (1970).

39. Byer, R. L., Kildal, H., Feigelson, R. S., "CdGeAs$_2$—A New Nonlinear Crystal Phasematchable at 10.6 μm," *Appl. Phys. Lett.*, **19**, 237 (1971).

40. Chemla, D. S., Kupecek, P. J., Robertson, D. S., and Smith, R. C., "Silver Thiogallate, a New Material with Potential for I. R. Devices," *Opt. Commun.*, **3**, 29 (1971).

41. Chemla, D. S., "Dielectric Theory of Tetrahedral Solids: Application to Ternary Compounds with Chalcopyrite Structure," *Phys. Rev. Lett.*, **26**, 1441 (1971).

42. Chen, F. S., "Optically Induced Change of Refractive Indices in LiNbO$_3$ and LiTaO$_3$," *J. Appl. Phys.*, **40**, 3389 (1969).

43. Chesler, R. B., Karr, M. A., and Geusic, J. E., "Repetitively Q-Switched Nd:YAlG-LiIO$_3$ 0.53 μ Harmonic Source," *J. Appl. Phys.*, **41**, 4125 (1970).

44. Collins, S. A., "Analysis of Optical Resonators Involving Focusing Elements," *Appl. Opt.*, **3**, 1263 (1964).

45. Comley, J., and Garmire, E., "Second-Harmonic Generation From Short Pulses," *Appl. Phys. Lett.*, **12**, 7 (1968).

46. Dana, E. S., and Ford, W. E., *A Textbook of Mineralogy*, 4th ed., New York: Wiley, 1951.

47. Day, G. W., "Linear and Nonlinear Optical Properties of Trigonal Selenium," *Appl. Phys. Lett.*, **18**, 347 (1971).

48. Dowley, M. W., "Efficient CW Second Harmonic Generation to 2573 Å," *Appl. Phys. Lett.*, **13**, 395 (1968).

49. Ducuing, J., and Bloembergen, N., "Doubling of Ruby in GaAs Surface Wave," *Phys. Rev. Lett.*, **10**, 474 (1962).

50. Ducuing, J., and Bloembergen, N., "Static Fluctuations in Nonlinear Optical Processes," *Phys. Rev.*, **133**, A1493 (1964).

51. Fay, H., Alford, W. J., and Dess, H. M., "Dependence of Second-Harmonic Phase-Matching Temperature in LiNbO$_3$ Crystals on Melt Composition," *Appl. Phys. Lett.*, **12**, 89 (1968).

52. Feynman, R. P., Leighton, R. B., and Sands, M., *The Feynman Lectures on Physics*, Vol. 3. Reading, Mass.: Addison-Wesley, 1966, Chapter IV.

53. Firester, A. H., "Parametric Image Conversion: Part I," *J. Appl. Phys.*, **40**, 4842 (1969); "Holography and Parametric Image Conversion: Part II," **40**, 4849 (1969); "Image Up-Conversion: Part III," **41**, 703, (1970).

54. François, G. E., "CW Measurement of the Optical Nonlinearity of Ammonium Dihydrogen Phosphate," *Phys. Rev.*, **143**, 597 (1966).

55. Franken, P. A., Hill, A. E., Peters, C. W., and Weinreich, G., "Generation of Optical Harmonics," *Phys. Rev. Lett.*, **7**, 118 (1961).

56. Gandrud, W. B., and Boyd, G. D., "Photomultiplier Detection of 10.6-μ Radiation Using CW Sum mixing in Ag_3SbS_3," *Opt. Commun.*, **1**, 187 (1969).

57. Gandrud, W. B., Boyd, G. D., and McFee, J. H., and Wehmeier, F. H., "Nonlinear Optical Properties of Ag_3SbS_3," *Appl. Phys. Lett.*, **16**, 59 (1970).

58. Gandrud, W. B., and Abrams, R. L., "Reduction in SHG Efficiency in Tellurium by Photo-Induced Carriers," *Appl. Phys. Lett.*, **17**, 302 (1970).

59. Gampel, L., and Johnson, F. M., "Index of Refraction of Single Crystal Selenium," *J. Opt. Soc. Am.*, **59**, 72 (1969).

60. Gampel, L., and Johnson, F. M., "IR Image Detection by CW Parametric Up-Conversion to the Visible," *IEEE J. Quant. Electr.*, **QE-4**, 354 (1968).

61. Garrett, C. G. B., and Robinson, F. N. H., "Miller's Phenomenological Rule for Computing Nonlinear Susceptibilities," *IEEE J. Quant. Electr.*, **QE-2**, 328 (1966).

62. Geusic, J. E., Levinstein, H. J., Rubin, J. J., Singh, S., and van Uitert, L. G., "The Nonlinear Optical Properties of $Ba_2NaNb_5O_{15}$," *Appl. Phys. Lett.*, **11**, 269 (1967).

63. Geusic, J. E., Levinstein, H. J., Singh, S., Smith, R. G., and van Uitert, L. G., "Continuous 0.532-μ Solid-State Source Using $Ba_2NaNb_5O_{15}$," *Appl. Phys. Lett.*, **12**, 306 (1968).

64. Giallorenzi, T. G., and Tang, C. L., "Quantum Theory of Spontaneous Parametric Scattering of Intense Light," *Phys. Rev.*, **166**, 225 (1968).

65. Giordmaine, J. A., "Mixing of Light Beams in Crystals," *Phys. Rev. Lett.*, **8**, 19 (1962).

66. Giordmaine, J. A., and Miller, R. C., "Tunable Coherent Parametric Oscillation in $LiNbO_3$ at Optical Frequencies," *Phys. Rev. Lett.*, **14**, 973 (1965).

67. Giordmaine, J. A., and Miller, R. C., "Optical Parametric Oscillation in the Visible Spectrum," *Appl. Phys. Lett.*, **9**, 298 (1966).

68. Giordmaine, J. A., and Miller, R. C., "Optical Parametric Oscillation in $LiNbO_3$," *The Physics of Quantum Electronics*, P. L. Kelley, B. Lax, and P. E. Tannenwald, Eds. New York: McGraw-Hill, 1966, p. 31.

69. Gordon, J. P., "Quantum Effects in Communications Systems," *Proc. IRE*, **50**, 1898 (1962).

70. Hagen, W. F., and Magnante, P. G., "Efficient Second-Harmonic Generation with Diffraction Limited and High-Spectral-Radiance Nd-Glass Lasers," *J. Appl. Phys.*, **40**, 219 (1969).

71. Harris, S. E., "Proposed Backward Wave Oscillation in the Infrared," *Appl. Phys. Lett.*, **9**, 114 (1966).

72. Harris, S. E., "Threshold of Multimode Parametric Oscillators," *IEEE J. Quant. Electr.*, **QE-2**, 701 (1966).

73. Harris, S. E., Oshman, M. K., and Byer, R. L., "Observation of Tunable Optical Parametric Fluorescence," *Phys. Rev. Lett.*, **18**, 732 (1967).

73a. Harris, S. E., "Tunable Parametric Oscillators," *Proc. IEEE*, **57**, 2096 (1969).

74. Heilmeyer, G. H., Ockman, N., Braunstein, R., and Kramer, D. A., "Relation Between Optical Second-Harmonic Generation and the Electrooptic Effect in the Molecular Crystal Hexamine," *Appl. Phys. Lett.*, **5**, 229 (1964).

75. Ho, I. T., and Siegman, A. E., "Passive Phase-Distortionless Parametric Limiting with Varactor Diodes," *IRE Trans. Microwave Theory Technol.* **MTT-9**, 459 (1961).

76. Hobden, M. V., Robertson, D. S., Davies, P. H., Hulme, K. F., Warner, J., and Midwinter, J. E., "Properties of Phase-Matchable Nonlinear Optical Crystal; Potassium Dithionate," *Phys. Lett.*, **22**, 65 (1966).

77. Hobden, M. V., and Warner, J., "The Temperature Dependence of the Refractive Indices of Pure Lithium Niobate," *Phys. Lett.*, **22**, 243 (1966).

78. Hobden, M. V., "Phase-Matched Second Harmonic Generation in Biaxial Crystals," *J. Appl. Phys.*, **38**, 4365 (1967).

79. Hobden, M. V., "The Dispersion of the Refractive Indices of Proustite (Ag_3AsS_3)," *Opto-Electronics*, **1**, 159 (1969).

80. Hulme, K. F., Jones, O., Davies, P. H., and Hobden, M. V. "Synthetic Proustite (Ag_3AsS_3): A New Crystal for Optical Mixing," *Appl. Phys. Lett.*, **10**, 133 (1967).

81. Izrailenko, A. N., Orlov, R. Yu., and Kopsik, V. A. "Ammonium Oxalate—A New Nonlinear Optical Material," *Sov. Phys. Crystallogr.* (Engl. transl.), **13**, 136 (1968).

82. Jenkins, F. A., and White H. A., *Fundamentals of Optics*, 3rd ed. New York: McGraw-Hill, 1957.

83. Jerphagnon, J., Batifol, E., Tsoucaris, G., and Sourbe, M., "Génération de Second Harmonique dans le Cinabre," *Compt. Rend.*, **265B**, 495 (1967).

84. Jerphagnon, J., "Optical Nonlinear Susceptibilities of Lithium Iodate," *Appl. Phys. Lett.*, **16**, 298 (1970).

85. Jerphagnon, J., and Kurtz, S. K., "Optical Nonlinear Susceptibilities: Accurate Relative Values for Quartz, Ammonium Dihydrogen Phosphate and Potassium Dihydrogen Phosphate," *Phys. Rev.*, **B1**, 1739 (1970).

86. Jerphagnon, J., "Optical SHG in Isocyclic and Heterocyclic Organic Compounds," *IEEE J. Quant. Electr.*, **QE-7**, 42 (1971).

87. Jerphagnon, J., and Newkirk, H. W., "Optical Nonlinear Susceptibilities of Beryllium Oxide," *Appl. Phys. Lett.*, **18**, 245 (1971).

88. Jerphagnon, J., "Invariants of the Third-Rank Cartesian Tensor, Optical Nonlinear Susceptibilities," *Phys. Rev.*, **B2**, 1091 (1970).

89. For a detailed explanation of crystal symmetry see, for example, C. Kittel, *Introduction to Solid State Physics*, 4th ed. New York: Wiley, 1971.

90. Kleinman, D. A., "Nonlinear Dielectric Polarization in Optical Media," *Phys. Rev.*, **126**, 1977 (1962).

91. Kleinman, D. A., "Theory of Optical Parametric Noise," *Phys. Rev.*, **174**, 1027 (1968).

92. Kleinman, D. A., and Boyd, G. D., "Infrared Detection by Optical Mixing," *J. Appl. Phys.*, **40**, 546 (1969).

93. Klinger, Y., and Arams, F. R., "Infrared 10.6-Micron CW Up-Conversion in Proustite Using an Nd:YAG Laser Pump," Proc. *IEEE*, **57**, 1797 (1969).

94. Kreuzer, L. B., "Single and Multimode Oscillation of the Singly Resonant Optical Parametric Oscillator," *Proc. Joint. Conf. Lasers and Opto-Electronics*, London, 1969, p. 53.

95. Kurtz, S. K., and Perry, T. T.; "A Powder Technique for the Evaluation of Nonlinear Optical Materials," *J. Appl. Phys.*, **39**, 3798 (1968).

96. Kurtz, S. K., Perry, T. T., and Bergman, J. G., Jr., "Alpha-Iodic Acid: A Solution-Grown Crystal for Nonlinear Optical Studies and Applications," *Appl. Phys. Lett.*, **12**, 186 (1968).

97. Kurtz, S. K., "New Nonlinear Optical Materials," *IEEE J. Quant. Electr.*, **QE–4,** 578 (1968).

98. Larsen, E. S., and Berman, H., *The Microscopic Determination of Non-Opaque Minerals*, 2nd ed. Washington, D.C.: Government Printing Office, Geological Survey Bulletin 1934, p. 848.

99. Levine, B. F., "Electrodynamical Bond-Charge Calculations of Nonlinear Optical Materials," *Phys. Rev. Lett.*, **22,** 787 (1969).

100. Levine, B. F., "A New Contribution to the Nonlinear Susceptibility Arising from Unequal Atomic Radii," *Phys. Rev. Lett.*, **25,** 440 (1970).

101. Louisell, W. H., *Coupled Mode and Parametric Electronics.* New York: Wiley, 1960.

102. Louisell, W. H., *Coupled Mode and Parametric Electronics.* New York: Wiley, 1960, see Chapter 6.

103. See, for example, W. H. Louisell, *Radiation and Noise in Quantum Electronics.* New York: McGraw-Hill, 1964, p. 279.

104. Maier, M., Kaiser, W., and Giordmaine, J. A., "Intense Light Bursts in the Stimulated Raman Effect," *Phys. Rev. Lett.*, **17,** 1275 (1966).

105. Maker, P. D., Terhune, R. W., Nisenoff, M., and Savage, C. M., "Effects of Dispersion and Focusing on the Production of Optical Harmonics," *Phys. Rev. Lett.*, **8,** 21 (1962).

106. Manley, J. M., and Rowe, H. E., "General Energy in Nonlinear Reactances," *Proc. IRE*, **47,** 2115 (1959).

107. Marple, D. T. F., "Refractive Index of ZnSe, ZnTe, and CdTe," *J. Appl. Phys.*, **35,** 539 (1964).

108. Marple, D. T. F., "Refractive Index of GaAs," *J. Appl. Phys.*, **35,** 1251 (1964).

109. McFee, J. H., Boyd, G. D., and Schmidt, G. D., "Redetermination of the Nonlinear Optical Coefficients of Te and GaAs by Comparison with Ag_3SbS_3," *Appl. Phys. Lett.*, **17,** 57 (1970).

110. McGeoch, M. W., and Smith, R. C., "Optimum Second-Harmonic Generation in Lithium Niobate," *IEEE J. Quant. Electr.*, **QE–6,** 203 (1970).

111. Midwinter, J. E., "Assessment of Lithium Meta-Niobate for Nonlinear Optics," *Appl. Phys. Lett.*, **11,** 128 (1967).

112. Midwinter, J. E., and Warner, J., "Up-Conversion of Near Infra-Red to Visible Radiation in Lithium Meta-Niobate," *J. Appl. Phys.*, **38,** 519 (1967).

112a. Midwinter, J. E., "Lithium Niobate: Effects of Composition on the Refractive Indices and Optical Second Harmonic Generation," *J. Appl. Phys.*, **39,** 3033 (1968).

113. Midwinter, J. E., "Image Conversion from 1.6 μ to the Visible in Lithium Niobate," *Appl. Phys. Lett.*, **12,** 68 (1968).

113a. Midwinter, J. E., "Parametric Image Converters," *IEEE J. Quant. Electr.*, **QE–4,** 716 (1968).

114. Midwinter, J. E., and Zernike, F., "Note on Up-Converter Noise Performance," *IEEE J. Quant. Electr.*, **QE–5,** 130 (1969).

115. Midwinter, J. E., "Infrared Up-Conversion in Lithium-Niobate with Large Bandwidth and Solid Acceptance Angle," *Appl. Phys. Lett.*, **14,** 29 (1969).

116. Miller, R. C., "Optical Second Harmonic Generation in Piezoelectric Crystals," *Appl. Phys. Lett.*, **5,** 17 (1964).

117. Miller, R. C., Boyd, G. D., and Savage, A., "Nonlinear Optical Interactions in LiNbO$_3$ without Double Refraction," *Appl. Phys. Lett.*, **6**, 77 (1965).

118. Miller, R. C., and Nordland, W. A., "Relative Signs of Nonlinear Optical Coefficients of Polar Crystals," *Appl. Phys. Lett.*, **16**, 174 (1970) and erratum, *ibid.*, **16**, 372 (1970).

119. Miller, R. C., Nordland, W. A., Kobb, E. D., and Bond, W. L., "Nonlinear Optical Properties of Lithium Gallium Oxide," *J. Appl. Phys.*, **41**, 3008 (1970).

120. Miller, R. C., and Nordland, W. A., "Absolute Sign of SHG Coefficients of Piezoelectric Crystals," *Phys. Rev.*, **B2**, 4896 (1970).

121. Nash, F. R., Bergman, J. G., Boyd, G. D., and Turner, E. H., "Optical Nonlinearities in LiIO$_3$," *J. Appl. Phys.*, **40**, 5201 (1969).

122. Nash, F. R., Boyd, G. D., Sargeant, M., III, and Bridenbaugh, P. M., "Effect of Optical Inhomogeneities on Phase Matching in Nonlinear Crystals," *J. Appl. Phys.*, **41**, 2564 (1970).

123. Nassau, K., Levinstein, J. J., Loiacono, G. M., Abrahams, S. C., Reddy, J. M., Bernstein, J. L., and Hamilton, J., "Ferroelectric Lithium Niobate" (5 papers), *J. Phys. Chem. Solids*, **27**, 983–1026 (1966).

124. Nath, G., and Haussuhl, S., "Large Nonlinear Optical Coefficient and Phase-Matched Second Harmonic Generation in LiIO$_3$," *Appl. Phys. Lett.*, **14**, 154 (1969).

125. Nath, G., Mehmanesch, H., and Gsanger, M., "Efficient Conversion of a Ruby Laser to 0.347 μ in Low-Loss Lithium Iodate," *Appl. Phys. Lett.*, **17**, 286 (1970).

126. "Table of Wave Numbers," *Nat. Bur. Std. (U.S.)*, Monogr. **3**, May 16, 1960.

127. Nelson, D. F., and Lax, M., "Double Phase Matching of Acoustically Induced Optical Harmonic Generation," *Appl. Phys. Lett.*, **18**, 10 (1970).

128. Newkirk, H. W., Smith, D. K., and Kahn, J. S., "Synthetic Bromellite," *Am. Mineral.*, **51**, 141 (1966).

129. Niizekei, N., Yomada, T., and Toyoda, H., "Growth Ridges, Etched Hillocks, and Crystal Structure of Lithium Niobate," *Jap. J. Appl. Phys.*, **6**, 318 (1967).

130. Nye, J. F., *Physical Properties of Crystals*. Oxford: Clarendon Press, 1960.

131. Okada, M., and Ieiri, S., "Kleinman's Symmetry Relation in Nonlinear Optical Coefficients of LiIO$_3$," *Phys. Lett.*, **34A**, 63 (1971).

132. Orlov, R. Yu., "Hippuric Acid as a Source of Second Harmonic in the Optical Range," *Sov. Phys. Crystallogr.* (Engl. transl.), **11**, 410 (1966).

133. Patel, C. K. N., "Optical Harmonic Generation in the Infrared Using a CO$_2$ Laser," *Phys. Rev. Lett.*, **16**, 613 (1966).

134. Patel, C. K. N., and Van Tran, N., "Phase-Matched Nonlinear Interaction Between Circularly Polarized Waves," *Appl. Phys. Lett.*, **15**, 189 (1969).

135. Peterson, G. E., Glass, A. M., and Negran, T. J., "Control of the Susceptibility of Lithium Niobate to Laser Induced Refractive Index Changes," *Appl. Phys. Lett.*, **19**, 130 (1971).

136. Phillips, R., "Temperature Variation of the Index of Refraction of ADP, KDP, and Deuterated KDP," *J. Opt. Soc. Am.*, **56**, 629 (1966).

137. Polloni, R., and Svelto, O., "Optimum Coupling for Intracavity Second Harmonic Generation," *IEEE J. Quant. Electr.*, **QE–4**, 528 (1968).

138. Rabin, H., and Bey, P. P., "Phase Matching in Harmonic Generation Employing Optical Rotary Dispersion," *Phys. Rev.*, **156**, 1010 (1967).

139. Radhakrishan, T., "The Dispersion, Birefringence, and Optical Activity of Quartz," *Proc. Indian Acad. Sci.*, **25A**, 260 (1947).

140. Seitz, F., *The Modern Theory of Solids*, 1st ed. New York: McGraw-Hill, 1940, p. 633.

141. Siegman, A. E., "Nonlinear Optical Power Limiter," *Appl. Opt.*, **1**, 739 (1962).

142. Singh, S., Bonner, W. A., Potopowicz, J. R., and Van Uitert, L. G., "Nonlinear Optical Susceptibility of Lithium Formate Monohydrate," *Appl. Phys. Lett.*, **17**, 292 (1970).

143. Singh, S., Draegert, D. A., and Geusic, J. E., "Optical and Ferroelectric Properties of Barium Sodium Niobate," *Phys. Rev.*, **B2**, 2709 (1970).

144. Singh, S., Potopowicz, J. R., van Uitert, L. G., and Wemple, S. H., "Nonlinear Optical Properties of Hexagonal Silicon Carbide," *Appl. Phys. Lett.*, **19**, 53 (1971).

145. Smith, H. A., and Townes, C., "Frequency Conversion and Detection of Infrared Radiation," unpublished.

146. Smith, H. A., and Mahr, H., "An Infrared Detector for Astronomy Using Up-Conversion Techiques," presented at the International Quantum Electronics Conference, Kyoto, Japan, September 1970.

147. See, for example, R. A. Smith, F. E. Jones, and R. P. Chasmar, *The Detection and Measurement of Infra-Red Radiation*, 2nd ed. New York: Oxford University Press, 1968.

148. Smith, R. G., Geusic, J. E., Levinstein, H. J., Rubin, J. J., Singh, S., and van Uitert, L. G., "Continuous Optical Parametric Oscillation in $Ba_2Nb_5O_{15}$," *Appl. Phys. Lett.*, **12**, 308 (1968).

149. Smith, R. G., "Theory of Intracavity Optical Second-Harmonic Generation," *IEEE J. Quant. Electr.*, **QE-6**, 215 (1970).

150. Smith, R. G., "Effect of Index Inhomogeneities on Optical Second-Harmonic Generation," *J. Appl. Phys.*, **41**, 3014 (1970).

151. Smith, R. G., "Effects of Momentum Mismatch on Parametric Gain," *J. Appl. Phys.*, **41**, 4121 (1964).

152. Soreff, R. A., and Moos, H. W., "Optical Second-Harmonic Generation in ZnS–CdS and CdS–CdSe Alloys," *J. Appl. Phys.*, **35**, 2152 (1964).

153. Suvorov, V. S., and Sonin, A. S., "Nonlinear Optical Materials," *Sov. Phys. Crystallogr.* (Engl. transl.), **11**, 711 (1967).

154. Suvorov, V. S., Sonin, A. S., and Rez, I. E., "Some Nonlinear Optical Properties of Crystals of the KDP Group," *Sov. Phys.—J. Expt. Theor. Phys.* (Engl. transl.), **26**, 33 (1968).

155. Tang, C. L., "Spontaneous Emission in the Frequency Up-Conversion Process in Nonlinear Optics," *Phys. Rev.*, **182**, 367 (1969).

156. Tien, P. K., Ulrich, R., and Martin, R. J., "Optical Second Harmonic Generation in Form of Coherent Čerenkov Radiation from a Thin-Film Waveguide," *Appl. Phys. Lett.*, **17**, 447 (1970).

157. Van Tran, N., Spalter, J., Manus, J., Ernest, J., and Kehl, D., "Generation of the Difference Frequency by Non-Collinear Light Beams in KDP Crystal," *Phys. Lett.*, **19**, 285 (1965).

158. Van Tran, N., Strnad, A. R., Jean-Louis, A. M., and Duraffourg, G., "Temperature-Dependent Phasematching for Far-Infrared Difference Frequency Generation in

InSb$_{1-x}$Bi$_x$ Alloy." *Proceedings of the Symposium on the Physics of Semimetals and Narrow Gap Semiconductors*, D. L. Carter and R. T. Bate, Eds., London: Pergamon, 1971.

159. Van Vechten, J. A., and Phillips, J. C., "New Set of Tetrahedral Covalent Radii," *Phys. Rev.*, **B2**, 2160 (1970).

160. Warner, J., "Phase-Matching for Optical Up-Conversion with Maximum Angular-Aperture, Theory and Practice," *Opto-Electronics*, **1**, 25 (1969).

161. Warner, J., "Photomultiplier Detection of 10.6-μ Radiation Using Optical Up-Conversion in Proustite," *Appl. Phys. Lett.*, **12**, 222 (1968).

162. See, for example, H. P. Weber and H. G. Danielmeyer, "Multimode Effects in Intensity Correlation Measurements," *Phys. Rev.*, **A2**, 2074 (1970).

163. Willard, G. W., "Use of the Etch Technique for Determining Orientation and Twinning in Quartz Crystals," *Bell Syst. Tech. J.*, **23**, 11 (1944).

164. Wincell, A. N., *Optical Properties of Organic Compounds*. New York: Academic Press, 1954.

165. Winchell, A. N., and Winchell, H., *Microscopical Characters of Artificial Inorganic Solid Substances*. New York: Academic Press, 1964.

166. Winchell, A. N., *Optical Properties of Minerals*. New York: Acdemic Press, 1965.

167. Wood, Elizabeth A., *Crystal Orientation Manual*. New York: Columbia University Press, 1963.

168. Wynne, J. J., and Bloembergen, N., "Measurement of the Lowest Order Nonlinear Susceptibility in III–V Semiconductors by Second-Harmonic Generation with a CO_2 Laser," *Phys. Rev.*, **188**, 1211 (1969).

169. Yarborough, J. M., and Amman, E. O., "Simultaneous Optical Parametric Oscillation, Second Harmonic Generation, and Difference-Frequency Generation," *Appl. Phys. Lett.*, **19**, 145 (1971).

169a. Yarborough, J. M., and Massey, G. A., "Efficient High Gain Parametric Generation in ADP Continuously Tunable Across the Visible Spectrum," *Appl. Phys. Lett.*, **18**, 438 (1971).

170. Yariv, A., *Quantum Electronics*. New York: Wiley, 1967, Appendix 5.

171. Young, J. F., Miles, R. B., Harris, S. E., and Wallace, R. W., "Pump Linewidth Requirements of the Optical Parametric Oscillator," *J. Appl. Phys.*, **42**, 497 (1971).

172. Zernike, F., "Refractive Indices of Ammonium Dihydrogen Phosphate and Potassium Dihydrogen Phosphate Between 2000 Å and 1.5 μ," *J. Opt. Soc. Am.*, **54**, 1215 (1964), and erratum, *ibid.*, **55**, 210 (1965).

173. Zernike, F., and Berman, P. R., "Generation of Far Infrared as a Difference Frequency," *Phys. Rev. Lett.*, **15**, 999 (1965), and erratum, *ibid.*, **16**, 177 (1966).

174. Zernike, F., "Temperature-Dependent Phase Matching of Far-Infrared Difference Frequency Generation in InSb," *Phys. Rev. Lett.*, **22**, 931 (1969).

175. Zernike, F., "Phasematched Far-Infrared Difference Frequency Generation," *Bull. Am. Phys. Soc. II*, **14**, 741 (1969).

Index